A Bridge to Higher Mathematics

The goal of this unique text is to provide an "experience" that would facilitate a better transition for mathematics majors to the advanced proof-based courses required for their major.

If you feel like you *love mathematics but hate proofs*, this book is for you. The change from example-based courses such as Introductory Calculus to the proof-based courses in the major is often abrupt, and some students are left with the unpleasant feeling that a subject they loved has turned into material they find hard to understand.

The book exposes students and readers to some fundamental content and essential methods of constructing mathematical proofs in the context of four main courses required for the mathematics major – probability, linear algebra, real analysis, and abstract algebra.

Following an optional foundational chapter on background material, four short chapters, each focusing on a particular course, provide a slow-paced but rigorous introduction. Students get a preview of the discipline, its focus, language, mathematical objects of interest, and methods of proof commonly used in the field. The organization of the book helps to focus on the specific methods of proof and main ideas that will be emphasized in each of the courses.

The text may also be used as a review tool at the end of each course and for readers who want to learn the language and scope of the broad disciplines of linear algebra, abstract algebra, real analysis, and probability, before transitioning to these courses.

Textbooks in Mathematics

Series editors:
Al Boggess, Kenneth H. Rosen

Classical Vector Algebra
Vladimir Lepetic

Introduction to Number Theory
Mark Hunacek

Probability and Statistics for Engineering and the Sciences with Modeling using R
William P. Fox and Rodney X. Sturdivant

Computational Optimization: Success in Practice
Vladislav Bukshtynov

Computational Linear Algebra: with Applications and MATLAB® Computations
Robert E. White

Linear Algebra With Machine Learning and Data
Crista Arangala

Discrete Mathematics with Coding
Hugo D. Junghenn

Applied Mathematics for Scientists and Engineers
Youssef N. Raffoul

Graphs and Digraphs, Seventh Edition
Gary Chartrand, Heather Jordon, Vincent Vatter and Ping Zhang

An Introduction to Optimization with Applications in Data Analytics and Machine Learning
Jeffrey Paul Wheeler

Encounters with Chaos and Fractals, Third Edition
Denny Gulick and Jeff Ford

Differential Calculus in Several Variables
A Learning-by-Doing Approach
Marius Ghergu

Taking the "Oof!" out of Proofs
A Primer on Mathematical Proofs
Alexandr Draganov

Vector Calculus
Steven G. Krantz and Harold Parks

Intuitive Axiomatic Set Theory
José Luis García

Fundamentals of Abstract Algebra
Mark J. DeBonis

A Bridge to Higher Mathematics
James R. Kirkwood and Raina S. Robeva

https://www.routledge.com/Textbooks-in-Mathematics/book-series/CANDHTEXBOOMTH

A Bridge to Higher Mathematics

Authored by
James R. Kirkwood
and
Raina S. Robeva

CRC Press
Taylor & Francis Group
Boca Raton London New York

CRC Press is an imprint of the
Taylor & Francis Group, an **informa** business

A CHAPMAN & HALL BOOK

Designed cover image: by Viwat Udompitisup/Shutterstock

First edition published 2024
by CRC Press
2385 Executive Center Drive, Suite 320, Boca Raton, FL 33431

and by CRC Press
4 Park Square, Milton Park, Abingdon, Oxon, OX14 4RN

CRC Press is an imprint of Taylor & Francis Group, LLC

© 2024 James R. Kirkwood and Raina S. Robeva

Library of Congress Cataloging-in-Publication Data

Names: Kirkwood, James R., author. | Robeva, Raina S., author.
Title: A bridge to higher mathematics / authored by James R. Kirkwood and
Raina S. Robeva.
Description: First edition. | Boca Raton, FL : CRC Press, 2024. | Series:
Textbooks in mathematics | Includes bibliographical references and
index.
Identifiers: LCCN 2023053045 | ISBN 9781032623856 (hardback) | ISBN
9781032611846 (paperback) | ISBN 9781032623849 (ebook)
Subjects: LCSH: Proof theory--Textbooks.
Classification: LCC QA9.54 .K568 2024 | DDC 511.3/6--dc23/eng/20240209
LC record available at https://lccn.loc.gov/2023053045

ISBN: 978-1-032-62385-6 (hbk)
ISBN: 978-1-032-61184-6 (pbk)
ISBN: 978-1-032-62384-9 (ebk)

DOI: 10.1201/9781032623849

Typeset in Nimbus Roman
by KnowledgeWorks Global Ltd.

Publisher's note: This book has been prepared from camera-ready copy provided by the authors.

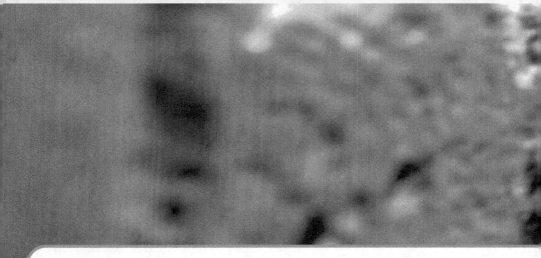

Contents

PREFACE

The undergraduate mathematics curriculum can roughly be described as having two components – one that emphasizes computations (such as Calculus) and the other that emphasizes theory (such as Abstract Algebra and Real Analysis). Typically, the computational courses are taken mainly, although not exclusively, in the first two years, while the theoretical courses are taken during the next two years. Many students – even the stronger ones – find that the transition does not occur without difficulty, which leaves many of them with the feeling that a subject they used to love has turned into material they find hard to understand and difficult to follow. As a result, students often choose to drop out from these proof-based courses and even leave the mathematics major. This makes retention in mathematics challenging at a time when strengthening, growing, and diversifying our STEM force has been identified as a national priority[1].

In the standard curriculum, introduction to abstract mathematics generally happens in the introduction to proofs courses such as Discrete Mathematics, although some institutions cover roughly the same material in courses titled Methods of Proof, Introduction to Proofs, or others with similar names. Such courses include elements from set theory, formal logic, methods of proof, number theory, combinatorics, relations, graph theory, and mathematical induction. In some cases, they also introduce matrix arithmetic and functions. These courses play an important role in the transition process from example-based to theory-based courses, but there are some weaknesses.

First, the time lag between when the course is taken and when the ideas are applied can be substantial. Also, which ideas apply to which future courses may not be obvious. The potpourri flavor of these foundational courses usually precludes a grand ending; that is, there is no obvious culmination that integrates the course topics. Because of this, students often fail to realize that the goal of the foundational course is to expose them to the fundamental nature and principles of constructing mathematical proofs and, thus, perceive these courses as something tangential to the main requirements for the major.

The second, and perhaps even more important weakness, is that such courses illustrate methods of proof (direct proof, proof by contraposition, proof by contradiction, proof by mathematical induction, etc.) almost exclusively by focusing on questions stemming from elementary number theory. This is understandable, as at this early stage of their education, students do not have a

---

[1] See, e.g., President's Council of Advisors on Science and Technology (PCAST). "Recommendations for strengthening American leadership in industries of the future," (2020).

broader mathematical base to build on. However, at this early stage of their careers, students lack the maturity to separate the methods of proof from the context in which the methods are employed. When proofs are constructed in their linear algebra and real analysis courses, for example, new content overshadows the logical structure. They generally feel that the proofs they see in these courses bear little resemblance to those included in their foundational course. Methods of proof students may have seen and applied in the context of, say, integer divisibility and modular arithmetic, would often appear as entirely new constructs when they are being used to prove linear independence in vector spaces, convergence of infinite sequences, existence of cluster points, continuity of functions, and so on. Further, even when finite groups and fields are introduced in abstract algebra courses, our experience shows that many students fail to see a connection with what they have learned about modular arithmetic in their introduction to proofs course out of context. The large number of new abstract structures they see in modern algebra, together with the need to understand the importance of working with their exact mathematical definitions when constructing a proof, makes the task difficult.

This text provides a slow-paced but rigorous introduction to probability, real analysis, linear algebra, and abstract (modern) algebra. We begin with a foundational chapter that gives an essential background in mathematical logic and methods of proof that students will use in the subsequent chapters. Each of the following chapters provides a "big picture" introduction of the course to which it pertains. The main goal is to bridge the transition between the "introduction to proofs" course and the theoretical courses for the mathematics majors by giving a preview of the discipline, its focus, language, mathematical object of interest, and common methods of proof. *We feel strongly that mathematics is learned in layers and have found that seeing this material for a second time often results in substantial improvement of student comprehension.* With this in mind, we believe that this book will be helpful to students who are considering a mathematics major and are not exactly sure what these higher-level courses are about and to instructors teaching such courses. Thus, Chapters 2–5 are written in a format that we believe will be most helpful to students when they take a standard course in the discipline. For example, the chapters on probability and linear algebra balance theory with computational approaches, while the chapters on real analysis and abstract algebra focus mainly on proofs.

Chapter 1 provides the necessary foundation in set theory, mathematical logic, and methods of proof needed for the later chapters. The material on sets is largely independent from the section devoted to general methods of proof, and there is a separate section devoted to proofs involving sets. In that later section, we also include some optional material on inclusion or equality for sets arising as images of functions and their inverses, which students may find useful when transitioning to a standard real analysis course. We should stress, however, that *Chapter 1 is not meant to be covered by instructors in its entirety*, unless the instructor is teaching a course in discrete mathematics or introduction to proofs and chooses to adopt the chapter as required reading for that course. For instructors teaching probability, real analysis, linear algebra, or abstract algebra, we recommend that they use "nuggets" from the chapter in a way that best fits the course structure and the pedagogy they employ.

Chapter 2 introduces discrete probability. We chose to present the use of conditional probability early in the context of using the multiplication principle in problems "without replacement," in contrast with using strictly counting techniques such as combinations or permutations. Our students have generally found this approach to make more sense, especially in cases when determining the number of combinations or permutations may be confusing to them. We have adopted the same "common sense" approach when it comes to the Bayes theorem, stating the respective theoretical result after the reader has already seen many examples, in which using basic principles for computing probabilities have been prioritized over applying the Bayes formula.

Chapter 3 is a gentle introduction to real analysis. It covers sequences of real numbers and their properties, bounds for sets of real numbers, infinite series, limits of functions, and continuous functions. We follow a traditional approach to the subject and provide a lot of examples with detailed step-by-step explanations and emphasis on the main ideas. We chose not to cover subsequences, as we believe that students who have mastered the included material will be able to transition to subsequences in their traditional courses without difficulties.

Chapter 4 exposes the reader to the language and methods of linear algebra. Here we give the general definition of vector space, then take a computational approach to solving systems of linear equations and to studying vectors in $\mathbb{R}^m$. We present several fundamental problems that link the solution set of a system of linear equations to its column space. We use the vector form of a system of linear equations to determine when a system has only the trivial solution and to study conditions under which the columns of a matrix span the entire space $\mathbb{R}^m$. In this context, we discuss when a set of vectors is a basis for $\mathbb{R}^m$. We end with a brief discussion of how answering these questions translates to important properties of linear transformations.

Chapter 5 is devoted to abstract algebra, where we cover the basics of group theory. We provide multiple examples of groups and subgroups, among which is the symmetric group, introduced through the symmetries of an equilateral triangle. Equivalence relations are presented as a background to cosets, which then serve to introduce normal groups and quotient subgroups. The focus in the chapter is on proofs.

We envision the material in this book being used in two ways. The first would be to offer a separate course at the beginning of a student's study of abstract mathematics (probably the second semester of the sophomore year) and select to cover two of Chapters 2–5, in addition to relevant material from the foundational Chapter 1. In some cases, this material may be integrated with some of the standard material in discrete mathematics or introduction to proofs courses. The second way would be to take the first couple of weeks at the beginning of a traditional proof-based course to study the chapter that pertains to that particular course. With each of these approaches come distinct challenges.

With the first approach, if students wait until, say, the second semester of their senior year to take real analysis, they will likely have forgotten what they studied two years before. In other words, we have a problem similar to what we identified for the traditional courses in discrete mathematics or introduction to proofs. With the second approach, the advantage is that the material for the specific subject will be fresh in the students' minds. The disadvantage is that the first couple of weeks of the semester are used to give an extended preview of the main course, and, of course, time is valuable. This may result in less material being covered.

Our belief is the optimal return will be achieved by the second approach, even though it would indeed shorten the time for the regular course. The advantage would be that introducing some of the main concepts, structures, and ideas during these first two to three weeks will help students see them as somewhat familiar when they are covered again (and, in many cases, more rigorously) later in the course. This takes us back to how learning abstract mathematics in "layers" helps one to better understand and retain the most important elements of the course. We believe that prioritizing quality of comprehension over quantity of covered material is a savvy strategy in the long run. Thus, seeing some of the same material presented from different angles throughout the course is likely to improve students' understanding and their learning experience.

Additionally, we envision this material being helpful to students who have completed regular courses in probability, real analysis, linear algebra, and abstract algebra as a review tool at the end of the course, e.g., in preparation for the final exam. Having taught these courses many times, we have noticed that students often "can't see the forest for the trees" – they may have mastered a lot of "techniques" throughout the course, but may still be unable to see how the concepts tie together to form a cohesive theory. Reading the relevant chapter at the end of their course would likely help them to better see important connections that may have been obscured by technical details when each of the concepts was first introduced.

Finally, for readers who want to learn the basic language and scope of the broad disciplines of probability, real analysis, linear algebra, and abstract algebra, our book can serve as a detailed preview to what these disciplines are about and to the mathematical questions they study. In such way, it will be of interest to high school students, mathematics teachers, researchers from non-STEM fields, and the general audience.

We would like to thank Katy Ott (Bates College) and Sloan Despeau (Western Carolina University) for supporting this project from its inception and providing valuable comments and advice. We are grateful to our editor, Bob Ross, who gave us suggestions and comments that helped us improve the manuscript. We are also grateful to our students at Sweet Briar College and Randolph-Macon College who, as our main audience over the years, have helped us fine-tune the pedagogy that motivated us to write this book. Finally, we thank our spouses, Bessie Kirkwood and Boris Kovatchev, for their patience and support throughout the process.

James Kirkwood, Raina Robeva
January 5, 2024

ABOUT THE AUTHORS

James R. Kirkwood holds a PhD in mathematics from the University of Virginia. He has authored and co-authored 20 published mathematics textbooks on various topics including calculus, real analysis, mathematical biology, and mathematical physics, many published by CRC Press. His original research was in mathematical physics, and he co-authored the seminal paper on a topic now called Kirkwood-Thomas Theory. He is the recipient of many awards for his teaching and research, including the Outstanding Faculty Award of the State Council of Higher Education for Virginia—the Commonwealth's highest honor for faculty at Virginia's public and private colleges and universities.

Raina S. Robeva is a professor of mathematics at Randolph-Macon College in Virginia. She holds a PhD in mathematics from the University of Virginia and is the lead author/editor of several textbooks and volumes in mathematical biology. She has led numerous educational and professional development initiatives at the interface of mathematics and biology sponsored by NSF, NIH, and MAA among others. Robeva is the founding Editor-in-Chief of Frontiers in Systems Biology, an open-access journal in the portfolio of Frontiers publications. In 2014, she was awarded the Outstanding Faculty Award of the State Council of Higher Education for Virginia "for her superior accomplishments in teaching, research, and public service."

1. Mathematical Logic and Methods of Proof

The study of abstract mathematics is primarily involved with proving theorems. In this section we introduce some fundamental mathematical objects and properties, give the rules of logic, and discuss common methods of proving theorems. This should enhance your understanding of the proofs given in subsequent chapters and provide a structure for you to construct your own proofs.

.1 Sets

The most basic idea of abstract mathematics is that of a set. In this section we introduce the mathematical notation and define some common operations on sets.

.1 Some Definitions and Notation

Sets are usually denoted by capital letters from the English or Greek alphabets, such as A, B, C, Ω, or Θ. A set is made up of elements (members) that can be of any nature, e.g., numbers, points (say, in a plane), symbols, or functions. A set must be well defined; that is, given a set A and an element p, exactly one of the following is true: Either p is a member of A (denoted $p \in A$) or p is not a member of A (denoted $p \notin A$). A common way to describe a set that has few members is to list the members between braces. For example, $A = \{3, 4, 5\}$ or $\Omega = \{\clubsuit, \Diamond, \heartsuit, \spadesuit\}$.

DOI: 10.1201/9781032623849-1

The order of elements presented within the curly brackets does not matter. Two sets are equal when they contain the same elements. For example, $\{3,4,5\} = \{5,3,4\}$ and $\{\clubsuit, \diamondsuit, \heartsuit, \spadesuit\} = \{\heartsuit, \clubsuit, \diamondsuit, \spadesuit\}$. This is in contrast for points in the plane where $(2,3) \neq (3,2)$. Notice that we use parentheses in the latter case and not curly brackets.

Repeats of elements don't matter, and each element is provided only once. For example, $\{2,6,*,2,*\}$ is the same set as $\{2,6,*\}$.

Some sets have so many elements and appear so often they have special symbols. For example, the set of all real numbers is denoted by $\mathbb{R}$.

Besides listing the elements of a set, an alternative way of defining a set is to give a condition that elements in the set, and only those elements, satisfy. The notation for this has the form $\{x \in A \mid P\}$ where P is the condition that elements of a larger set A must satisfy. For example, $\{x \in \mathbb{R} \mid x^2 = 4\} = \{-2,2\}$.

If $a, b \in \mathbb{R}$, we use the following notation to denote intervals of real numbers:

- Closed intervals, denoted $[a,b] = \{x \in \mathbb{R} \mid a \leq x \leq b\}$;
- Open intervals, denoted $(a,b) = \{x \in \mathbb{R} \mid a < x < b\}$;
- Half-open intervals, denoted
  - ❖ $(a,b] = \{x \in \mathbb{R} \mid a < x \leq b\}$;
  - ❖ $[a,b) = \{x \in \mathbb{R} \mid a \leq x < b\}$;
- Infinite intervals, denoted
  - ❖ $(a,\infty) = \{x \in \mathbb{R} \mid a < x\}$;
  - ❖ $[a,\infty) = \{x \in \mathbb{R} \mid a \leq x\}$;
  - ❖ $(-\infty,b) = \{x \in \mathbb{R} \mid x < b\}$;
  - ❖ $(-\infty,b] = \{x \in \mathbb{R} \mid x \leq b\}$.

In addition to using $\mathbb{R}$ for the set of real numbers, other common sets of numbers and the notation we use to denote them, are:

$\mathbb{N} = \{1,2,3,....\}$ – the set of natural numbers;

$\mathbb{Z} = \{..., -4,-3,-2,-1,0,1,2,3,...\}$ – the set of integers;

$\mathbb{Q} = \{\frac{p}{q} \mid p \text{ and } q \text{ are integers}, q \neq 0\}$ – the set of rational numbers;

$\mathbb{R} = \{x \mid x \in (-\infty,\infty)\}$ – the set of all real numbers.

Sometimes, it is useful to create new sets from known ones. The Cartesian product of sets is one example.

Definition 1.1.1 The *Cartesian product* of two sets A and B, denoted $A \times B$, is defined as

$$A \times B = \{(a,b) \mid a \in A, b \in B\}.$$

That is, the set $A \times B$ is the set of *ordered pairs* (a,b), where the first element comes from the set A and the second from the set B.

■ **Example 1.1** Let $A = \{5, \spadesuit, \blacksquare\}$ and $B = \{a, f\}$. Then

$$A \times B = \{(5,a), (5,f), (\spadesuit,a), (\spadesuit,f), (\blacksquare,a), (\blacksquare,f)\}.$$

■

The definition generalizes naturally to a Cartesian product of any finite number of sets. For example, the Cartesian product $A \times B \times C$ of three sets $A, B,$ and C is defined as the set of all ordered triples (a,b,c), where $a \in A$, $b \in B$, and $c \in C$. One special example is the Cartesian power of a single set.

Definition 1.1.2 Given a set A and a positive integer n, the *n-th Cartesian power of A*, denoted A^n is the set

$$A^n = \{(x_1, x_2, \ldots, x_n) \mid x_1 \in A, x_2 \in A, \ldots, x_n \in A\}.$$

■ **Example 1.2** Let A be a set of two elements: $A = \{H, T\}$, where H (heads) and T(tails) represent the possible outcomes from tossing a coin. Then, the possible outcomes from tossing a coin seven times in a row can be described with the Cartesian power A^7 of sequences of length 7 comprised from Hs and Ts. So (H,T,H,H,T,T,T) is one of the elements of A^7. ■

Suppose we have a set B and form a new set A by taking some of the elements of B. Then A is a subset of B. More formally, we have the following definition.

Definition 1.1.3 Suppose A and B are sets. If every element of A is also an element of B we say that A is a *subset* of B and we write $A \subseteq B$. When this is not the case, we write $A \nsubseteq B$. We also say that B *contains* A or that A *is contained in* B. When A is contained in B, but $A \neq B$, we say that A is a *proper subset of* B. When it is necessary to underscore that, we write $A \subset B$.

■ **Example 1.3** Let $A = \{1, 2, 3\}$, $B = \{1, 7, *\}$, and $C = \{1, *, 5, 7, 2\}$. Since every element of B is also in C, we have that B is a subset of C. Since the two sets are not the same, we write $B \subset C$. The set A has 3 as an element, and $3 \notin C$, so A is not a subset of C. We write $A \not\subset C$. The elements of A are positive integers, so we have that $A \subset \mathbb{N}$. The sets B and C contain the element $*$, which is not an integer, so $B \not\subseteq \mathbb{N}$ and $C \not\subseteq \mathbb{N}$.   ∎

The set $\{\}$, that is, the set that has no elements, is called the *empty set*. It is customary to denote this set by the symbol $\varnothing$. The empty set is considered to be a subset of any set.

Let's spend some time to understand why the empty set is considered to be a subset of any set. Definition 1.1.3 tells us that $A \subseteq B$, if every element of A is also in B. Thus, to have that $A \not\subseteq B$, we must be able to find an element $a \in A$, such that $a \notin B$. The empty set $\varnothing$ does not have any elements, so, given a set B, we won't be able to find such an element of $\varnothing$. Thus, we have to conclude that $\varnothing \subseteq B$, for any set B.

∎ **Example 1.4** Let $S = \{*, a, 4\}$. List all subsets of S.

One usually begins with the empty set, then lists all one-element subsets, followed by all two-element subsets, and ending with the set itself. So, we have

$$\varnothing, \{*\}, \{a\}, \{4\}, \{*, a\}, \{*, 4\}, \{a, 4\}, \{*, a, 4\}.$$

∎

∎ **Example 1.5** Find all subsets of $B = \{(1, 2, 3)\}$. We have to be careful here about understanding the notation well. Since the parentheses in $(1, 2, 3)$ indicate this is an ordered triple, our set B contains a single element, which is the ordered triple $(1, 2, 3)$. Thus, there are only two subsets of B – the empty set $\varnothing$ and the set B itself.   ∎

1.1.2 Operations on Sets

Just as numbers can be combined with operations such as addition, subtraction, multiplication, and division, there are various operations that can be performed on sets. Suppose A and B are sets. The following operations are commonly used:

- The *union* of A and B, denoted $A \cup B$, is the set that combines the elements of A with the elements of B. That is,

$$A \cup B = \{x \mid x \in A \text{ or } x \in B\};$$

- The *intersection* of A and B, denoted $A \cap B$, is the set of elements that A and B have in common. That is,

$$A \cap B = \{x \mid x \in A \text{ and } x \in B\};$$

- The *set difference*, denoted $A - B$, is the set that contains all elements of A that are not in B. That is,

$$A - B = \{x \mid x \in A \text{ and } x \notin B\}.$$

Another set operation is called the *complement of a set*. The definition requires the idea of a universal set, which we now discuss.

When dealing with a set, we can almost always regard it as a subset of some larger set, appropriately chosen in the context of what is being discussed. That is, there are many such sets, but we choose one pertinent to our setting. This larger set we choose is the *universal set* for our problem.

Consider as an example, the set of prime numbers $P = \{2, 3, 5, 7, 11, 13, \dots\}$. Since each prime number is a positive integer, it would make sense to think of P as a subset of the positive integers $\mathbb{N}$ and ignore any other sets, the elements of which would be irrelevant to our discussion. So, we will select $\mathbb{N}$ to be the universal set we will work with.

Almost every useful set in mathematics can be regarded as having some natural universal set. For instance, the unit circle in the plane is the set of points

$$C = \{(x, y) \in \mathbb{R}^2 \mid x^2 + y^2 = 1\}.$$

Since all these points are in the plane $\mathbb{R}^2$, it is natural to regard $\mathbb{R}^2$ as a universal set for C.

In the absence of specifics, if A is a set, a universal set that contains A is often denoted as U. The set U is also called *the universe of A*. We are now ready to define the complement operation.

- The *complement* of a set A, denoted by A^c is the set

$$A^c = U - A.$$

■ **Example 1.6** Let $U = \{x \in \mathbb{Z} \mid -10 \leq x \leq 10\}$, and let

$$A = \{-1, 2, -5, 1\}, \quad B = \{1, 8, -5\}, \quad C = \{1, 7\}, \quad D = \{6, -2\}.$$

With this, we have

$$A \cap B = \{-5, 1\};$$
$$C \cup D = \{1, 7, 6, -2\};$$
$$D^c = \{-10, -9, -8, -7, -6, -5, -4, -3, -1, 0, 1, 2, 3, 4,$$
$$5, 7, 8, 9, 10\};$$
$$B \times C = \{(1, 1), (1, 7), (8, 1), (8, 7), (-5, 1), (-5, 7)\};$$
$$C \times B = \{(1, 1), (1, 8), (1, -5), (7, 1), (7, 8), (7, -5)\};$$
$$B \cap D = \varnothing;$$
$$D^c - A = \{-10, -9, -8, -7, -6, -4, -3, 0, 3, 4, 5, 7, 8,$$
$$9, 10\};$$
$$B \times C - C \times B = \{(1, 7), (8, 1), (8, 7), (-5, 1), (-5, 7)\}.$$

Notice that, in general, the order of sets in the Cartesian product of two sets matters. Namely, $B \times C \neq C \times B$, in general. Consequently, in this example, $B \times C - C \times B \neq \varnothing$. ∎

In thinking about sets, it is sometimes helpful to draw informal, schematic diagrams of them. In doing this we often represent a set with a circle, which we regard as enclosing all the elements of the set. Such graphical representations of sets are called *Venn diagrams*, after the British logician John Venn (1834–1923) who first introduced them. They can illustrate how sets combine using various operations and suggest some rules that may hold universally for operations on sets.

For two sets A and B (see Figure 1.1), the common elements are represented by the points in the plane where the sets overlap and the respective set differences are depicted as points which are in one of the sets but not in

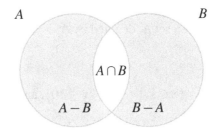

Figure 1.1: Venn diagram for the intersection $A \cap B$ (white area where A and B overlap) and the set differences $B - A$ and $A - B$.

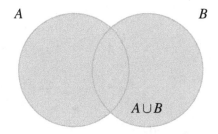

Figure 1.2: Venn diagram for the union $A \cup B$.

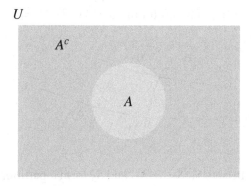

Figure 1.3: A Venn diagram for a set A and its complement A^c within the universe U.

the other. Note that just as when subtracting numbers $A - B$ is, in general, different from $B - A$. Figure 1.2 depicts the union $A \cup B$.

A Venn diagram depicting A and A^c within a universal set U is given in Figure 1.3.

Venn diagrams can be drawn for any number of sets and provide insight for stating some rules that hold for set operations. For example, the shaded region in Figure 1.4 depicts the set $(A \cup B) \cap C$. However, the same visualization is obtained for the set $(A \cap C) \cup (B \cap C)$, which suggests the following rule may hold:

$$(A \cup B) \cap C = (A \cap C) \cup (B \cap C).$$

Indeed, in Section 1.4.2, we will prove that this rule holds for any sets A, B, and C.

Sometimes it is helpful to "break" a set A into non-overlapping pieces as in Figure 1.5, where no two pieces have elements in common, and, when the

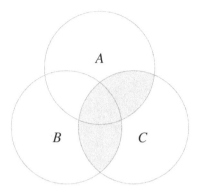

Figure 1.4: The Venn diagram for the set $(A \cup B) \cap C = (A \cap C) \cup (B \cap C)$.

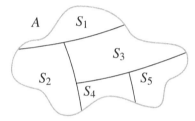

Figure 1.5: The sets S_1, S_2, S_3, S_4, and S_5 form a partition of the set A.

pieces are put together, they produce the entire set A. This visual description leads to the formal definition of a set partition.

Definition 1.1.4 Given a set A, we say that the sets $S_1, S_2, \ldots, S_n$ form a *partition of* A, if the following two conditions are satisfied:
1. $A = S_1 \cup S_2 \cup \cdots \cup S_n$;
2. For any two different sets S_i and S_j, we have $S_i \cap S_j = \varnothing$.
We also say that A is *partitioned into* sets $S_1, S_2, \ldots, S_n$.

■ **Example 1.7** Consider the set of all integers $\mathbb{Z}$ and the sets S_1, containing all odd numbers and the set S_2, containing all even numbers. Since $\mathbb{Z} = S_1 \cup S_2$ and $S_1 \cap S_2 = \varnothing$, S_1 and S_2 form a partition of $\mathbb{Z}$. ■

■ **Example 1.8** Let A be the set containing the 52 cards in a standard deck. Let S_1 be the set of all clubs, S_2 be the set of all diamonds, S_3 be the set of all hearts, and S_4 be the set of all spades in the deck. Then, the sets S_1, S_2, S_3, and S_4 form a partition of A. ■

Exercise 1.1 List all subsets for each of the sets

$$A = \{(1,3),(-2,-1)\}, \; B = \{1,3,-2,-1\}.$$

Exercise 1.2 Consider again the sets from Exercise 1.1. Find
1. $A \cap B$;
2. $A \cup B$.

Exercise 1.3 Consider the universal set

$$U = \{1,2,3,4,5,6,7,8\}$$

and the sets

$$A = \{2,8,5,7\}, \; B = \{3,1,8,5\}, \; C = \{1,2,4,7\}, \; D = \{6,1\}, \; E = \{3\}.$$

For each of the sets below, (i) Draw the associated Venn diagram and (ii) List its elements. Use proper notation.
1. $A \cup D$;
2. $B \cap C$;
3. $A - B$;
4. $E \cap D^c$;
5. $B - A$;
6. $B \cup C^c$;
7. $(A \cup D)^c$;
8. $(A \cap D)^c$;
9. $(A \cup D^c) \cap B$;
10. $(A \cap B) \cup E$.

Exercise 1.4 Let A, B, C be the following sets:

$$A = \{x \in \mathbb{R} : x \in (1,3]\}, \quad B = \{x \in \mathbb{R} : x \in (-4,1]\}, \quad C = \{x \in \mathbb{R} : x \in (-\infty,3)\}.$$

Find the sets below.
1. $A \cap C$;
2. $B \cup A$;
3. $A \cap B$;
4. $B \cap C$;
5. $A \cup C$;
6. $B^c - A$;
7. $A^c \cap B$.

Exercise 1.5 Give three different partitions of the set $A = \{1, 3, 5, 7, 9, 11, 13, 15, 17, 19\}$ by listing the elements of each of the sets forming the partition.

1.2 Mathematical Logic

1.2.1 Propositions

Mathematical logic provides a system of rules that allows us to study propositions (statements) and the relations between them in a rigorous way. This system of reasoning lays the foundation for a proper construction of mathematical proofs. We begin by introducing some terminology.

Definition 1.2.1 A *proposition* is a declarative sentence for which we can ask whether what it states is true or false in a given context. The term *statement* is also used, and we will use "statement" and "proposition" interchangeably.

■ **Example 1.9** The following are propositions. We can ask if each is true or false for a specific time/place.
- I am a mathematics major.
- My shirt is green.
- Today is a hot day.
- The sun shines at night.
- Robert's cat is named Charlie.
- $2 + 2 = 5$.

The following are not propositions for the reason stated in parentheses.
- Are you OK? (This is a question, not a declarative sentence);
- Go home. (This is a directive; cannot reasonably ask if what it states is true or false);

- Perhaps you should reconsider your position. (This makes a suggestion, doesn't state anything; cannot ask if it is true or false.)

∎

In mathematical logic, propositions are often denoted by lower case letters such as p, q, r, so we can write, e.g.,

$p =$ Today is Monday.

Depending on whether a proposition p is true or false, we will often say that p has value True (T) or value False (F). Notice that some statements are always true or always false, independent of the circumstances. For instance, "$2 + 2 = 5$" is always false and "January is the first month of the year" is always true. Other statements, like "My shoes are black," may change their truth value depending on the circumstances. Thus, we may view a proposition as a *logical variable* with two possible values T and F. Such variables are also called *Boolean variables* after the English mathematician George Boole (1815–1864) who noticed that logical variables can be combined with logical operations in a way that establishes an analogy with algebraic operations. We turn to this next.

1.2.2   The AND and OR Connectives

Given two or more propositions, we can form longer (compound) statements using the logical connectives AND and OR. So, let

$p =$ Today the outside temperature is 74^oF, and

$q =$ It is cloudy outside.

We can form the compound statement

$r =$ Today the outside temperature is 74^oF AND it is cloudy outside.

Symbolically, we write

$$r = p \wedge q, \tag{1.1}$$

where we use $\wedge$ to indicate AND. We read Equation (1.1) as "r is equal to p and q." In a similar way, we can use an OR to combine any two statements, using $\vee$ to denote this operation. Now the statement

$$s = p \vee q, \tag{1.2}$$

which is read "s is equal to p or q," is the statement

$s =$ Today the outside temperature is 74°F OR it is cloudy outside.

Notice how the symbol ∧ we use for the logical AND is similar to the symbol ∩ that we use to refer to the intersection of two sets: $A \cap B$ (recall that $A \cap B$ is defined as all points that are in A AND B). The symbol ∨ looks similar to the symbol ∪ denoting the union of two sets: $A \cup B$ (recall that $A \cup B$ is defined as all points that are in A OR in B). This observation could serve as a hint for how to remember the new notation more easily.

With the new operations, Equations (1.1) and (1.2) look like algebraic expressions. And just as with any algebraic expressions, we would like to know their logical value when we know the values of the logical variables p and q.

Recall how in grade school you learned the multiplication table for multiplying two integers $a \cdot b$, given the values of a and b. In the same way, Table 1.1 shows the values for the operations AND (∧) and OR (∨). Such tables are called *truth tables*.

Note that these tables are arranged differently from the familiar format of the multiplication table. Since each variable can only take two values, T and F, there are only four possible combinations for the values of p and q ($p =$ T, $q =$ T; $p =$ T, $q =$ F; $p =$ F, $q =$ T; and $p =$ F, $q =$ F). Each row of the tables represents one of those combinations, and the result is given in the third column.

p	q	$p \wedge q$
T	T	T
T	F	F
F	T	F
F	F	F

p	q	$p \vee q$
T	T	T
T	F	T
F	T	T
F	F	F

Table 1.1: Truth tables for the logical operations AND (left panel) and OR (right panel).

Notice also that the AND (∧) operation is True only when both logical variables have value True and is False otherwise. The OR (∨) operation is True when at least one of the logical variables has value True and is False only when both variables have value False (i.e., in mathematical logic, we use the *inclusive* OR). This is different from how most of us use OR in everyday life in the "either... or" sense, the exclusive sense (one or the other but not both). There are several reasons for using the inclusive OR in mathematical logic, one of which, as we will see soon, is that the OR and AND operators are now linked in a nice way by the so-called De Morgan's laws.

2.3 The Negation Operator

Given any proposition p, we can form its negation by stating that p is not true.

Consider the examples shown in Table 1.2. In a spoken or written language, we have various ways of expressing the negation of a statement, but in mathematics, we think of the negation of a statement p as NOT p (it is not true that p happens).

Proposition p	Negation of p
The sky is blue.	It is not true that the sky is blue. (The sky is not blue.)
$2 + 7 = 9$.	It is not true that $2 + 7 = 9$. ($2 + 7 \neq 9$.)
2023 is a leap year.	It is not true that 2023 is a leap year. (2023 is not a leap year.)

Table 1.2: Negating propositions. The statements in parentheses is how we would likely express the negation in a general conversation.

The notation for negation, the so-called *negation operator*, is $\sim$. Given a statement p, its negation is denoted $\sim p$. When $p = $ T, we have $\sim p = $ F, and when $p = $ F, we have $\sim p = $ T (see Table 1.3).

p	$\sim p$
T	F
F	T

Table 1.3: The truth table for the negation operator.

Notice that if we take the negation of the negation, in other words $\sim (\sim p)$, we will get back the value of p we started with, and we write

$$\sim (\sim p) = p.$$

What we mean by this is that the values of the left-hand side and the right-hand side are the same for any given value of p. This is perhaps the simplest example of *logical equivalence*, at which we will look more closely soon.

1.2.4 The If-Then Construct

Another way to connect two propositions is to use the If-Then construct. Let

$$p = \text{It is raining.} \tag{1.3}$$

$$q = \text{The shingles on the roof of my house are wet.} \tag{1.4}$$

We can now form a new statement s:

$s = $ If it is raining, then

       the shingles on the roof of my house are wet.

We use the symbol $\Rightarrow$ for the If-Then construct and write

$$s = p \Rightarrow q.$$

This is read *If p, then q*.

We now want to know what the truth value of s is, based on the values of p and q. The four possible cases are presented in Table 1.4.

p	q	$p \Rightarrow q$
T	T	T
T	F	F
F	T	T
F	F	T

Table 1.4: The truth table for the If-Then construct. Note that $p \Rightarrow q$ is false only when p is true and q is false.

Let's pause to examine Table 1.4 and see if we can make sense of what it says when we use the statements p and q from (1.3) and (1.4).

- If both p and q are true, the statement s will certainly be true (unless your house is under a giant umbrella that always keeps your roof dry, which is unlikely).
- If p is true and q is false, the statement s can be read as

       If it is raining, then

          the shingles on the roof of my house are not wet.

This statement would be false (because, most likely, your house is not under a giant umbrella that always keeps your roof dry).

- The last two rows of the table may look strange at first, but think about it this way. What we are saying there is that if p is false, then anything *can* happen: If it is not raining, the rooftop of your house may be wet ($q = \text{T}$), if, for example, it has stopped raining a few minutes ago, or it may be dry ($q = \text{F}$), which would be the most likely scenario, if it hasn't rained for a while. Either of these can happen, so either of these could be true.

There is one very important observation we should make here that leads to a widely used terminology in mathematics. Table 1.4 tells us that if p is true, then the only case in which $p \Rightarrow q$ is true is when q is also true. In other words, if p is true, q *necessarily* must be true too, for the If-Then statement to be true. In mathematical language, we also often say that q is *a necessary condition* for p. In the same context, since p being true implies that q is true, we say that p is a *sufficient condition* for q. We summarize this in the following definition.

Definition 1.2.2 When p is true and $p \Rightarrow q$ is true, we say that p is a *sufficient condition* for q and that q is a *necessary condition* for p.

Exercise 1.6 Write the following propositions in symbolic form $p \Rightarrow q$, clearly identifying the propositions p and q. If possible, identify if $p \Rightarrow q$ is true or false.
1. If I like the book I am reading, I will finish it in a week.
2. Leap years have 366 days.
3. Knowing that an integer p is prime, means that p is odd.
4. Having a cat implies that you are not a dog person.
5. Driving is not safe, if you cannot keep your eyes on the road.

2.5 Logical Equivalence

You know from high-school algebra that two expressions may look different, yet mean the same thing. For example, if x, y, z are any numbers, the distributive law tells us that

$$(x+y)z = xz + yz.$$

The expressions on the left- and right-hand side look different but they will produce the same values, no matter which values you pick for x, y, and z. The same concept can be applied to logical expressions.

Definition 1.2.3 We say that two logical expressions are *logically equiva-lent* if they produce the same values as outputs for the same input values. When two logical expressions s and t are logically equivalent, we will write $s = t$.

The best way to establish logical equivalence is by constructing the truth table of each expression and comparing the values for the same inputs. We saw one such example earlier:

■ **Example 1.10** For any statement p, we have

$$p = \sim(\sim p). \tag{1.5}$$

Let's prove this using a truth table. Since we already have Table 1.3 for the negation operator, we just need to add another column for $\sim(\sim p)$, in which we give the negation of the values in the second column (see Table 1.5). Since the values in the first and third columns match for each line, we have proved the logical equivalence from Equation (1.5).

p	$\sim p$	$\sim(\sim p)$
T	F	T
F	T	F

Table 1.5: The truth table that proves $p = \sim(\sim p)$.

■

Theorem 1.2.1 (*De Morgan's laws of logic*) For any two propositions p and q, the following logical equivalences hold.

$$\sim(p \wedge q) = (\sim p) \vee (\sim q) \tag{1.6}$$

and

$$\sim(p \vee q) = (\sim p) \wedge (\sim q). \tag{1.7}$$

Proof. Let's first decide whether what each law says makes sense, if we think about it. Recall that $p \wedge q$ is true only when both p and q are true. The negation $\sim(p \wedge q)$ tells us $p \wedge q$ is not true, which means p or q (or, possibly, both) are false. In symbols, this is expressed as $(\sim p) \vee (\sim q)$.

For the second law, recall that $p \vee q$ is false only when both p and q are false. In this case, $\sim(p \vee q)$ being true tells us that $p \vee q$ did not happen.

Which means that p did not happen (expressed as $\sim p$) AND q did not happen (expressed as $\sim q$). In symbols, we then write this as $(\sim p) \wedge (\sim q)$.

Let's now provide the mathematical proof, using truth tables. We will prove the equivalency given by Equation 1.6. The other law is proved the same way and is left as an exercise.

We start with two columns giving the four possible values for the Boolean variables p and q. Since we will also need their negations, we have added columns for $\sim p$ and $\sim q$. After that we need a column for each of $p \wedge q$, $\sim (p \wedge q)$, and $(\sim p) \vee (\sim q)$. The results are presented in Table 1.6. The last two columns show that the expressions $\sim (p \wedge q)$ and $(\sim p) \vee (\sim q)$ produce the same values for identical inputs. This proves that they are logically equivalent.

p	q	$\sim p$	$\sim q$	$p \wedge q$	$\sim (p \wedge q)$	$(\sim p) \vee (\sim q)$
T	T	F	F	T	F	F
T	F	F	T	F	T	T
F	T	T	F	F	T	T
F	F	T	T	F	T	T

Table 1.6: Proof of the De Morgan's laws from Equation (1.6).

■

We next present a logical equivalency that is of great importance when constructing proofs.

Definition 1.2.4 Given two propositions p and q, the statement $(\sim q) \Rightarrow (\sim p)$ is called the *contrapositive statement* of $p \Rightarrow q$.

Theorem 1.2.2 For any propositions p and q, the statement $p \Rightarrow q$ is logically equivalent to its contrapositive statement $(\sim q) \Rightarrow (\sim p)$.

Proof. The proof provides another opportunity to practice truth tables.

In addition to the columns for p and q, we need columns for $\sim p$, $\sim q$, $p \Rightarrow q$, and $(\sim q) \Rightarrow (\sim p)$. Computing these expressions independently from the others for all possible values of p and q gives Table 1.7. Since the last two columns are identical, we have proved that $p \Rightarrow q$ and $(\sim q) \Rightarrow (\sim p)$ are logically equivalent.

■

p	q	$\sim p$	$\sim q$	$p \Rightarrow q$	$(\sim q) \Rightarrow (\sim p)$
T	T	F	F	T	T
T	F	F	T	F	F
F	T	T	F	T	T
F	F	T	T	T	T

Table 1.7: A truth table showing that $p \Rightarrow q$ and $(\sim q) \Rightarrow (\sim p)$ are logically equivalent.

Exercise 1.7 Show that for any statement p, the statement $p \wedge \sim p$ is false. As we will see in Section 1.3.3, this fact is the basis for constructing proofs by contradiction.

Exercise 1.8 Prove the De Morgan's law from Equation (1.7) using a truth table.

Exercise 1.9 Show that the statement $q \Rightarrow p$ is *not* logically equivalent to $p \Rightarrow q$. The statement $q \Rightarrow p$ is called the *converse* of $p \Rightarrow q$.

Exercise 1.10 Show that $\sim p \Rightarrow q \wedge (\sim q)$ is logically equivalent to p.

Exercise 1.11 Are the expressions below logically equivalent? Answer by constructing the appropriate truth tables.
1. $\sim (p \Rightarrow q)$ and $p \wedge (\sim q)$;
2. $p \vee (q \wedge r)$ and $(p \vee q) \wedge r$. *Hint.* In this case, there are three logical variables, each of which can have a value true or false. Thus, to answer the question, you will need to fill out Table 1.8.

1.2.6 The If-And-Only-If Construct

Having defined logical equivalence allows us to construct more complex statements from existing propositions. If p and q are propositions, the *if-and-only-if construct*, denoted by $\Leftrightarrow$ is defined as

$$p \Leftrightarrow q = (p \Rightarrow q) \wedge (q \Rightarrow p).$$

Knowing how to find the truth values of "if-then" statements, we can now construct a truth table for $p \Leftrightarrow q$ (see Table 1.9).

p	q	r	$q \wedge r$	$p \vee (q \wedge r)$	$p \vee q$	$(p \vee q) \wedge r$
T	T	T				
T	T	F				
T	F	T				
T	F	F				
F	T	T				
F	T	F				
F	F	T				
F	F	F				

Table 1.8: A truth table to be completed for Exercise 1.11.

p	q	$p \Rightarrow q$	$q \Rightarrow p$	$p \Leftrightarrow q = (p \Rightarrow q) \wedge (q \Rightarrow p)$
T	T	T	T	T
T	F	F	T	F
F	T	T	F	F
F	F	T	T	T

Table 1.9: A truth table for $p \Leftrightarrow q$. The statement $p \Leftrightarrow q$ is true only when *both* $p \Rightarrow q$ and $q \Rightarrow p$ are true.

Recall that when p is true and $p \Rightarrow q$ is true, we say that p is a sufficient condition for q and q is a necessary condition for p. If, at the same time, we know that when q is true and $q \Rightarrow p$ is true, this makes q a sufficient condition for p, and p a necessary condition for q. This leads to the following definition.

Definition 1.2.5 When $p \Leftrightarrow q$ is true, we say that p is a *necessary and sufficient condition* for q and q is a necessary and sufficient condition for p.

Exercise 1.12 Are the following statements logically equivalent?

1. $p \Rightarrow (q \wedge r)$ and $(p \wedge \sim q) \Rightarrow r$;

2. $p \Leftrightarrow q$ and $(\sim p) \vee q$;

3. $p \Leftrightarrow (q \vee r)$ and $q \Rightarrow p \wedge r$.   ∎

2.7   Universal and Existential Statements

When constructing mathematical proofs, two types of questions are of significant importance. The first question is whether we can show that all objects from a group have a specific property. Examples include:

- All integers are fractions with denominator equal to 1.
- For all negative integers a and b, the product ab is a positive integer.
- All numbers divisible by 4 are even numbers.
- For all right triangles with sides a, b and a hypotenuse c, $a^2 + b^2 = c^2$.

Such statements are called *universal statements*. They can be (re)phrased to begin with the words "All" or "For all," but we usually won't convey their meaning this way. The examples above may sound awkward because, in a conversation, we are a lot more likely to say something like:

- Integers are fractions with denominator equal to 1.
- The product ab of two negative integers is a positive integer.
- A number divisible by 4 is an even number.
- If a right triangle has sides a, b and a hypotenuse c, then $a^2 + b^2 = c^2$.

The second question is whether we can find an object that satisfies a certain property. Examples include:

- There exists a positive integer that is not a prime number.
- There exists a black tulip.
- There is an even number that is not divisible by 6.
- There exists a white cat.

Such statements are called *existential statements*. Existential statements can be (re)phrased to begin with the words "There exists" or "There is," although in a conversational format the above statements are more likely to sound like:

- Not all positive integers are prime numbers.
- A tulip can be black.
- An even number may not be divisible by 6.
- Some cats are white.

In this section we discuss how to translate universal and existential statements from conversational format to the more structured mathematical format.

We begin by formalizing the mathematical format. In the case of universal statements, we want to show that all elements from a certain group S have some property p. In the case of existential statements, we want to be able to find an element within a certain group that has some property q. Again, the group of objects S, within which we look, varies depending on context. It may be the set of all positive integers, the set of all tulip types, the set of negative real numbers, and so on.

Two mathematical symbols are often used in this context, mainly as a shorthand for expressing universal and existential statements in mathematical form:

- ∀ is the symbol that reads "For all";
- ∃ is the symbol that reads "There exists." Sometimes, it may make more sense to read it as "There is" or "We/one can find."

With this notation, we can express a universal statement as

$$\forall a \in S, p. \tag{1.8}$$

This reads "For all elements a in the group S, statement p is true."

An existential statement can be expressed as

$$\exists\, a \in S, q. \tag{1.9}$$

This reads "There exists an element a in the group S, for which statement q is true."

Note that *just as any other statements, existential and universal statements may be true or false.*

■ **Example 1.11** Identify each statement as universal, existential or neither. Write any universal and existential statement, in the format provided by expressions (1.8) and (1.9), identifying the set S and the propositions p and q.

1. The statement

   There are male calico cats.

   is an existential statement. The set S is the set of all calico cats. The proposition q is "This is a male cat." Now $\exists\, a \in S, q$ reads "There exists a calico cat that is a male cat."

2. The statement

   Today is Sunday.

   is neither an existential nor a universal statement.

3. The statement

   Cows eat grass.

   is a universal statement. The set S is the set of cows. The proposition p is "This cow eats grass." Now $\forall a \in S, p$ can be read as "All cows eat grass" or as "If a is a cow, then a eats grass."

4. The statement

   I don't have classes on Sundays.

   is a universal statement. Here S is the set of all Sundays, and the statement p can be phrased as "I don't have classes." Now $\forall a \in S, p$ can be read as "If it's Sunday, then I don't have classes."

■

Exercise 1.13 Classify each of the statements below as universal, existential, or neither.

1. Birds eat worms.

2. The sky is purple.

3. Some days in the summer feel longer than they are.

4. Every family has problems.

5. The set of integers in denoted by $\mathbb{Z}$.

6. There are no bad people.

7. All roses are red.

8. Costco stores do not accept Mastercard.

9. Summer days in Virginia are hot and humid.

10. Everybody in the village was scared.

Exercise 1.14 Write each statement that you identified as an existential or universal in Exercise 1.13 in symbolic form as in Equations (1.8) and (1.9). Be sure to describe what the set S is and what the properties (propositions) p or q are.

1.2.8 Negating Universal and Existential Statements

Forming the negation of a given statement is often important when constructing proofs. In Section 1.2.5, we learned how to negate AND and OR statements using the De Morgan's laws. We now discuss how to negate universal and existential statements.

A universal statement claims that *all* elements in a set S have a certain property. If this is not the case, we should be able to find at least one element a of S, which *does not* have that property. Consider the universal statement discussed in Example 1.11:

All cows eat grass.

In symbols (see again Example 1.11 for the notation), we wrote this as

$$\forall a \in S, \ p.$$

To say that this is *not* the case would mean to find a cow that does not eat grass. More formally, we would then say,

There exists a cow that does not eat grass.

This is an existential statement! So, in symbols, we would write

$$\exists\, a \in S,\ \sim p.$$

This example illustrates the general rule for negating universal statements, represented by the following logical equivalence:

$$\sim (\forall a \in S,\ p) = \exists\, a \in S,\ \sim p. \tag{1.10}$$

Don't let the notation here intimidate you! The above equation states a simple fact. If someone claims that *every* element of a set S satisfies some property, all you need to do to prove them wrong is to show them *one* element of the set S, which *does not* have that property. As we will see in the next section, this approach is used when we want to prove that a universal statement is false.

Let's now move to negating an existential statement. In Example 1.11 we also considered the statement

There are male calico cats.

In symbols (using again the notation introduced in Example 1.11), we wrote

$$\exists\, a \in S,\ q.$$

Now, what would it mean for this statement *not* to be true? This would mean that no matter which calico cat you examine, it won't be male. This would lead to the universal statement

All calico cats are female.

Symbolically, we write

$$\forall a \in S,\ \sim q.$$

This example illustrates the following logical equivalency:

$$\sim (\exists\, a \in S,\ q) = \forall a \in S,\ \sim q. \tag{1.11}$$

Again, let's emphasize the meaning behind this notation. Let's say someone claims they *can find* an element from a set S with a certain property, and you want to prove them wrong. To do so, you will need to show that *no* element of S has that property. In other words, you will have to check that *no matter which* (any) element of S you chose, it *won't* have that property. Take some time to understand the logic here. Just as we said before, we don't always use "any" or "for all" to make universal statements in everyday conversations, but the formalism would make more and more sense as you gain experience.

■ **Example 1.12** Let L be a given real number and $S \subset \mathbb{R}$. Negate the statement:

> For any number $\varepsilon > 0$,
>
> the interval $(L - \varepsilon, L + \varepsilon)$ contains infinitely many elements of S.

This is a universal statement about all positive real numbers ε with the property "the interval $(L - \varepsilon, L + \varepsilon)$ contains infinitely many elements of S." The negation should say that we can find a real number ε for which the property fails. Thus, we have the negation:

> There is an $\varepsilon > 0$, such that the
>
> interval $(L - \varepsilon, L + \varepsilon)$ does *not* contain infinitely many elements of S.

Equivalently, we could also phrase the negated statement as follows:

> There is an $\varepsilon > 0$, such that
>
> the interval $(L - \varepsilon, L + \varepsilon)$ contains only finitely many elements of S.

■

■ **Example 1.13** Let $f : X \to Y$ be a function. Negate the statement:

> For any $x_1, x_2 \in X$, for which $x_1 \neq x_2$, we have $f(x_1) \neq f(x_2)$.

This is a universal statement that we may rephrase as:

> For any $x_1, x_2 \in X$, with the property $x_1 \neq x_2$, we have $f(x_1) \neq f(x_2)$.

Thus, it's negation is the existential statement

> We can find $x_1, x_2 \in X$, with the property $x_1 \neq x_2$, for which
>
> $f(x_1) \neq f(x_2)$ is not true.

In a more streamlined form, we have the negation:

> We can find $x_1, x_2 \in X$, for which $x_1 \neq x_2$, but $f(x_1) = f(x_2)$.

■

In summary, Equations (1.10) and (1.11) give us two simple rules. The negation of a universal statement is an existential statement and the negation of an existential statement is a universal statement. There is a nice logical symmetry here, but we would caution against using memorization. Instead, when doing the exercises below, focus on the following:

- Learn how to recognize existential and universal statements.
- When you need to form a negation of such statements, ask yourself
  - What would it mean for a "for all" statement to not be true?
  - What would it mean for a "there exists" statement to not be true?
- Focus on how to formulate such negations using common sense and avoid thinking of Equations (1.10) and (1.11) as formulas that will produce the needed answers.

> **Exercise 1.15** Negate each statement. Express the negation in a way that would make it sound most natural in conversational or written English. It may be helpful to first write each statement in symbolic form as in Equations (1.8) or (1.9). Keep in mind that if a statement is true, its negation will be false and vice versa.
> 1. Children's books have pictures.
> 2. There are interesting places to see in France.
> 3. Every positive integer greater than 2 is a sum of two prime numbers.
> 4. Cats take naps during the day.
> 5. Some like their coffee cold.
> 6. Sunflowers face the sun during daytime.
> 7. There is an $\varepsilon > 0$, such that the intervals $(L - \varepsilon, L + \varepsilon)$ and $(K - \varepsilon, K + \varepsilon)$ are disjoint.
> 8. For all positive integers $n > N$, we have $|x_n - L| < \varepsilon$.
> 9. Every interval on the real line $\mathbb{R}$ contains an irrational number.
> 10. All differentiable functions are continuous.

2.9 Negating Compound universal and Existential Statements[1]

In Section 1.2.8 we showed how to negate universal and existential statements. Table 1.10 gives us a summary.

When constructing mathematical proofs, one often needs to negate more complex statements, in which the statements p and q from Table 1.10 are themselves universal or existential statements. In that case, as our next examples shows, we simply have to apply the same rules multiple times. These examples are of particular importance in real analysis.

---

[1]This material may be omitted on the first reading and revisited when limits of sequences and functions are discussed in Chapter 3.

	Statement	Negation
Universal	$\forall a \in S, p$	$\exists a \in S, \sim p$
Existential	$\exists a \in S, q$	$\forall a \in S, \sim q$

Table 1.10: Rules for negating universal and existential statements. Here p and q state the property that the element $a \in S$ satisfies.

■ **Example 1.14** Let $f : X \to Y$ be a function. Negate the statement:

$s =$ For every $y \in Y$, we can find $x \in X$, for which $f(x) = y$.

We begin by realizing that s is a universal statement with condition p, and that p is an existential statement with property q, as marked below.

$$\text{For every } y \in Y, \underbrace{\text{we can find } x \in X, \text{ for which } \overbrace{f(x) = y.}^{q}}_{p}$$

So, the negation of the universal statement s is

$$\sim s = \exists y \in Y, \sim p. \tag{1.12}$$

and the negation of the existential statement p is

$$\sim p = \forall x \in X, \sim q. \tag{1.13}$$

The negation of q is

$$f(x) \neq y. \tag{1.14}$$

Now, we obtain the negation of s, starting at the line numbered (1.12), and reading it (or writing it in words, not symbols). Move to the line (1.13) when you need $\sim p$, then to the line (1.14) when you need $\sim q$. With this, we get:

$\sim s =$ There is a $y \in Y$, such that for all $x \in X$, we have $f(x) \neq y$.

■

This "iterative" approach to negating existential and universal statements can be applied to even more complex statements. In our next example the statement s gives the definition of a function f being continuous at a point x_0. In Chapter 3, you will understand the meaning behind this definition. For now, just consider it to be a compound statement that you have to negate using the same approach as in Example 1.14.

■ **Example 1.15** Let $f : \mathbb{R} \to \mathbb{R}$ be a function and x_0 be a fixed point within its domain. Negate the following statement s:

$$s = \text{For any } \varepsilon > 0, \text{ there exists a } \delta > 0, \text{ such that, if } |x - x_0| < \delta,$$
$$\text{then } |f(x) - f(x_0)| < \varepsilon. \tag{1.15}$$

Let's first break the universal compound statement s into simpler ones as marked below.

For any $\varepsilon > 0$, there exists a $\delta > 0$, such that if $|x - x_0| < \delta$, then $\underbrace{|f(x) - f(x_0)| < \varepsilon}_{r}$.

$\underbrace{}_{q}$

$\underbrace{}_{p}$

Notice that here p is an existential statement and q is a universal statement. With this, we have:

$$\sim s = \exists\, \varepsilon > 0, \sim p; \tag{1.16}$$
$$\sim p = \forall \delta > 0, \sim q;$$
$$\sim q = \exists\, x \text{ with } |x - x_0| < \delta, \sim r;$$
$$\sim r = |f(x) - f(x_0)| \geq \varepsilon.$$

Now, starting from line (1.16), start writing down (in words) the negation of s, moving to the following line when you need to write $\sim p$, and so on. With this, we now have the negation of statement s, as desired:

There is an $\varepsilon > 0$, such that for any $\delta > 0$,
we can find an x with $|x - x_0| \leq \delta$, for which $|f(x) - f(x_0)| \geq \varepsilon$.

■

Exercise 1.16 Follow the same approach as in Examples 1.14 and 1.15 to state the negation of each statement. Assume here that $x_n, n = 1, 2, 3, \ldots,$ is a given list (sequence) of real numbers, where x_1 is its first element, x_2, the second, and so on. In Chapter 3 we will formalize this in the context of infinite sequences of real numbers.

1. There exists a real function f that is discontinuous at every point $x \in \mathbb{R}$.

2. For every $\varepsilon > 0$, there exists a number $N > 0$, such that for all $n > N$, we have $|x_n - L| < \varepsilon$. (Here, assume that L is a given fixed real number.)

3. For every number M, there exists a number N, such that for all $n > N$, $x_n > M$.

4. For every number K, there exists a number N, such that for all $n > N$, $x_n < K$.

5. There is a function $f : \mathbb{R} \to \mathbb{R}$ that is continuous everywhere but differentiable nowhere.

1.3 Methods of Proof

We said earlier that universal and existential statements, as any statements, can be true or false. But how do we decide which one is which? A decision of this type is called a proof. There are several common techniques that we discuss here. Our list is by no means comprehensive, but it covers the basic types of proof and the logical rationale behind them.

1.3.1 Proving Existential Statements True

An existential statement of the form

$$\exists\, a \in S, q,$$

is most commonly proved to be true by finding an example for an element of the set S that satisfies the given property q. To find such an example may be easy or difficult, depending on the specific problem. For instance, the statement

"Not all odd numbers are prime."

can be proved by giving the example that $15 = 3 \cdot 5$ is an odd number, but it is not a prime number. Thus, there exists an odd number that is not prime. Of course, there are many other examples (e.g., 21, 25, and so on) that prove the same statement to be true, but we only need one.

On the other hand, to prove that

"There is a positive integer with 250 digits that is a product of exactly two prime numbers."

may take considerably longer. That claim was proved to be true in 2020 (see, e.g., https://en.wikipedia.org/wiki/RSA_numbers) by identifying the number known as RSA-250, and its two prime factors. The example was found in response to the Factoring Challenge issued by the security company

RSA in 1991 to encourage research in computational number theory, so it took a long time to find such a number. It was shown that:

$$
\begin{aligned}
\text{RSA-250} =&\,2140324650240744961264423072839333563008614715144755017799775 \\
&\,49208814180234471401366433455190958046796109928518724709145 8 \\
&\,7687396261921557363047454770520805119056493106687691590019 75 \\
&\,940569345745223058932597669747168173806936489469987157849497 \\
&\,5937497937 \\
=&\,(64135289477071580278790190170577389084825014742943447208116 8 \\
&\,596320245323446302386235987526683477087376619255856946397988 \\
&\,53367) \\
\times&\,(33372027594978156556226010605355114227940760344767554666784 5 \\
&\,2098702384172921003708025744867329688187756571898625803693 20 \\
&\,62711),
\end{aligned}
$$

where the two factors on the right can be checked to be prime numbers.

3.2   Proving Universal Statements False – Counterexamples

We can prove that a universal statement is false by producing a counterexample. Let's say you traveled to Scotland and saw a great number of white sheep. You may then be inclined to generalize your observation and formulate the universal statement

$p =$ All sheep in Scotland are white.

Proving this to be true may be difficult. No matter how many white sheep you have seen, you won't have a proof for this universal statement, until you have indeed seen every single sheep in Scotland and verified it has white fleece. On the other hand, seeing just one black sheep would immediately prove your claim false. An example that contradicts a universal claim is called a *counterexample*. We know from Section 1.2.8 that if the universal statement $\forall\, a \in S,\ p$ is false when its negation $\exists\, a \in S,\ \sim p$ is true and vice versa. Thus, if we can find a counterexample, we have proved the universal statement false.

Depending on the universal claim, finding a counter example to disprove it may be very easy or extremely difficult.

Let's consider the claim

"All prime numbers are odd numbers."

The number 2 is a prime number, and it's even. This is a counter example. We have thus proved that the claim "All prime numbers are odd numbers" is false.

Consider, on the other hand, the statement

> Every even natural number greater than 2 is the sum of two prime
> numbers.

This is a famous statement, the truth value of which is still unknown. What this means is that no one has found a counterexample yet, and no one has been able to prove it either. A claim of such nature is called a *conjecture*, and this is the well-known *Goldbach's conjecture*, after the mathematician Christian Goldbach (1690–1764) who first formulated it. It is one of the oldest unsolved problems in mathematics. Note that trying to prove the Goldbach's conjecture by diligently checking that *every* even natural number can be written as the sum of two primes would be impossible (even if we had a lifetime to devote to the task) because there are infinitely many such numbers. So how can we prove that a universal statement is true?

1.3.3 Proving Universal Statements True

There are several widely used methods to prove universal statements true. Getting enough experience to use them correctly takes time and careful applications of the rules of logic described in the previous sections. As you take more proof-oriented courses, you will get more practice and gain confidence. You will also gradually develop a good sense for the advantages of each approach and the conditions under which they may be most effective. In this section, we will introduce some terminology and provide selected examples that will get you started. Subsequent chapters will provide more examples.

Direct Proof

Recall that very often universal statements can be translated into If-Then constructs. Consider the following statement as an example:

> The median to the base of an isosceles triangle is perpendicular
> to the base.

We can reformulate this statement using If-Then as:

> If ABC is an isosceles triangle with $AB = BC = a$, the median AD
> to the base AC is perpendicular to the base (see Figure 1.6).

Thus, we need to prove

$$p \Rightarrow q,$$

where

    $p = ABC$ is an isosceles triangle with $AB = BC = a$, and

       D is a point on AB, such that $AD = DC$;

    $q =$ The median AD to the base AC of triangle ABC is perpendicular

       to the base.

    Recall also from Section 1.2.4 that $p \Rightarrow q$ is only false when p is true and q is false. Thus, to prove $p \Rightarrow q$ is true, we should show that when p is true, q is also true.

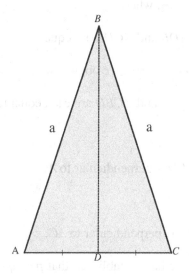

Figure 1.6: An isosceles triangle with $AB = BC = a$. The median to the base AC is perpendicular to the base.

    A direct proof begins with assuming that p is true, then establishes a sequence of true If-Then statements

$$p \Rightarrow c_1, \quad c_1 \Rightarrow c_2, \quad \ldots, c_{n-1} \Rightarrow c_n, \quad c_n \Rightarrow q, \tag{1.17}$$

where the statements $c_1, c_2, \ldots, c_n$ are to be proved (or have already been proved) to be true. The proof ends with concluding that q is true. Many proofs that you have seen in your geometry class in school are of this type. We will now illustrate this approach, emphasizing each step using the notation from line (1.17).

■ **Example 1.16** Prove that the median to the base of an isosceles triangle is perpendicular to the base.

Let ABC be an isosceles triangle with $AB = BC = a$ and D be the midpoint of the segment AC. Consider now the triangles ABD and CBD. Since $AD = DC$, $AB = BC$, and the BD is a side they have in common, the triangle are congruent. With

$c_1 = ABD$ and CBD are congruent triangles,

we thus have $p \Rightarrow c_1$.

Now that we have established that c_1 is true, what do we know about congruent triangles? We know that angels between corresponding sides are equal. Thus, we have $c_1 \Rightarrow c_2$, where

$c_2 = $ The angles $\angle ADB$ and $\angle CDB$ are equal.

Next, since $\angle ADB + \angle CDB = 180°$, we obtain that $c_2 \Rightarrow c_3$, where

$c_3 = $ The angles $\angle ADB$ and $\angle CBD$ are each equal to $90°$.

Thus, $c_3 \Rightarrow c_4$, where

$c_4 = $ The segment BD is perpendicular to AC.

Therefore $c_4 \Rightarrow q$ where

$q = $ The median BD is perpendicular to AC.

Thus, using direct proof, we have established that $p \Rightarrow q$ is true.

In practice, we would usually omit naming the intermediate statements, and the proof will be written as follows:

Let ABC be an isosceles triangle with $AB = BC = a$ and D be the midpoint of the segment AC. Since $AD = DC$, $AB = BC$, and the BD is a side they have in common, the triangles ADB and CDB are congruent. The angles $\angle ADB$ and $\angle CDB$ are now equal, as angles between corresponding sides of congruent triangles. Because D is on the line segment AC, we then have $\angle ADB + \angle CDB = 180°$, so $\angle ADB = \angle CDB = 90°$. Therefore $AD \perp AB$.   ∎

Many direct proofs presented at the introductory level, concern properties regarding even and odd integers.

Definition 1.3.1 An integer a is an *even integer*, if it can be written in the form $a = 2k$, where $k \in \mathbb{Z}$. An integer b is an *odd integer*, if it can be written as $b = 2m + 1$, where $m \in \mathbb{Z}$.

■ **Example 1.17** The sum of any two odd numbers is an even number.

As in our previous example, it is common to begin with introducing some notation to facilitate the proof.

Let a and b be two odd numbers. By definition, this means, we can write $a = 2n + 1$ and $b = 2m + 1$, where $n, m \in \mathbb{Z}$. Now

$$a + b = (2n + 1) + (2m + 1) = 2n + 2m + 2 = 2(n + m + 1) = 2k,$$

where $k \in \mathbb{Z}$ (since n, m, and 1 are integers). Therefore $a + b$ is even.   ■

■ **Example 1.18** Prove that if a is even and b is odd, then $a + b$ is odd.

Since a is even, we have that $a = 2m$, for some $m \in \mathbb{Z}$. Since b is odd, we have $b = 2n + 1$, for some $n \in \mathbb{Z}$. Now

$$a + b = 2m + (2n + 1) = 2(m + n) + 1 = 2k + 1, \text{ where } k = m + n \in \mathbb{Z}.$$

Therefore, by definition, $a + b$ is odd.   ■

■ **Example 1.19** Let $a = 3m + 1$ for some $m \in \mathbb{Z}$, and $b = 3n + 2$, for some $n \in \mathbb{Z}$. Prove that there is an integer k, for which

$$a + b = 3k.$$

We have

$$a + b = (3m + 1) + (3n + 2) = 3m + 3n + 3 = 3(m + n + 1) = 3k,$$
$$\text{where } k = m + n + 1 \in \mathbb{Z},$$

completing the proof.   ■

> **Exercise 1.17** Let $a = 4m + 3$ and $b = 4n + 1$. Prove that $a + b = 4k$, for some $k \in \mathbb{Z}$.   ■

> **Exercise 1.18** Prove that the sum of any two even numbers is an even number.   ■

> **Exercise 1.19** Let $a, b \in \mathbb{Z}$ be such that $a = bk$, for some $k \in \mathbb{Z}$ and $b = am$, for some $m \in \mathbb{Z}$. Prove that $a = b$ or $a = -b$.   ■

> **Exercise 1.20** Let $n \in \mathbb{Z}$. Prove that the product $n(n+1)$ is even. *Hint.* Consider two cases: (1) The number n is even, and (2) The number n is odd. ■

> **Exercise 1.21** Prove that the product of any two odd numbers is an odd number. ■

Indirect Proof (Proof by Contrapositive)

Indirect proofs are based on the logical equivalence of the statements $p \Rightarrow q$ and $\sim q \Rightarrow \sim p$ that we established in Section 1.2.5. Thus, if we want to show that $p \Rightarrow q$ is true, we can do this by establishing $\sim q \Rightarrow \sim p$. Depending on the problem, the latter may be easier to accomplish than a direct proof.

If we need to show $p \Rightarrow q$ by a contrapositive proof, we begin with the words "Let $\sim q$" (or something to that effect) and conclude the proof with "Therefore $\sim p$."

Below, we give two examples. The first proof can be done easily by using either a direct approach or a proof by contrapositive. The second example would be difficult to prove directly, but the contrapositive approach delivers it relatively easily.

■ **Example 1.20** If the number $7x+9$ is even, then the integer x is odd.

Here, with the understanding that x denotes an integer, we have

$$p = \text{The number } 7x+9 \text{ is even,}$$

and

$$q = \text{The number } x \text{ is odd.}$$

Direct Proof: Suppose $7x+9$ is even (Suppose p). Thus, by definition, $7x+9 = 2m$ for some integer m. Subtracting $6x+9$ from both sides, we get $x = 2m - 6x - 9$. Thus

$$x = 2m - 6x - 9 = 2m - 6x - 10 + 1 = 2(m - 3x - 5) + 1 = 2k + 1,$$

where $k = m - 3x - 5$ is an integer (since x, m, and 5 are integers). Therefore, x is odd (Therefore q).

Indirect Proof: Suppose x is even (Suppose $\sim q$). This means that $x = 2n$, for some integer n. Now

$$7x+9 = 7(2n) + 9 = 14n + 8 + 1 = 2(7n+4) + 1 = 2k + 1,$$

where $k = 7n + 4$ is an integer. Therefore, $7x+9$ is odd. (Therefore $\sim p$.) ■

Note that for the example above, we were able to carry out the direct proof because we obtained an expression for x relatively easily by subtracting $6x+9$ from both sides. Our next examples shows that this may be technically challenging in some cases, thus favoring a proof by contraposition.

■ **Example 1.21** If x is an integers and the number x^2 is odd, then x is odd.
   Here, we have

$$p = \text{The number } x^2 \text{ is odd,}$$

and

$$q = \text{The number } x \text{ is odd.}$$

A direct proof would not be easy here. If x^2 is odd, we know $x^2 = 2m+1$ for some integer m. However, expressing x from here would be problematic, as taking a square root from $2m+1$ may not be an integer.

   An indirect proof, however, works very well. We begin by assuming that x is even. (Assume $\sim q$). Then we can write $x = 2m$, for some integer m. Now

$$x^2 = (2m)^2 = 4m^2 = 2(2m^2) = 2k,$$

where $k = 2m^2$ is an integer (since m is an integer). Therefore x^2 is even. (Therefore $\sim p$.)   ■

Exercise 1.22 Prove that if $x \in \mathbb{Z}$ and x^2 is even, then x is even.   ■

Exercise 1.23 Let $m \in \mathbb{Z}$. Prove that if $m^2 - 2m + 9$ is odd, then m is even.   ■

Exercise 1.24 Prove that if x^2 is even, x cannot be written in the form $x = 4k + 1$, where $k \in \mathbb{Z}$.   ■

Exercise 1.25 Let $n \in \mathbb{Z}$. Prove that if n^2 cannot be written as $3k, k \in \mathbb{Z}$, then n cannot be written as $3m, m \in \mathbb{Z}$.   ■

Exercise 1.26 Let $m, n \in \mathbb{Z}$. If both nm and $n+m$ are even, then both n and m are even.
   Hint. Let $p = $ both nm and $n+m$ are even, and $q = $ both n and m are even. To carry out a contrapositive proof, assume $\sim q$. That is, assume that

n or m are odd. Let's say n is odd. Now prove $\sim p$ by considering two cases: Case (1) m is even, and Case (2) m is odd. ∎

Proof by Contradiction

Let's say we want to prove a statement of the form $p \Rightarrow q$. Recall that when p is true, the only way for $p \Rightarrow q$ to be false is when q is false. This is exactly how a proof by contradiction begins. To prove $p \Rightarrow q$ is true, we assume the opposite – that is, we assume $p \wedge (\sim q)$. Then we use definitions and logic to show that for some claim c, $c \wedge (\sim c)$ must be true, which is impossible (see Exercise 1.7). The impossibility arises from the assumption that $p \Rightarrow q$ is false. Our assumption then must be incorrect, thus proving that $p \Rightarrow q$ is true.

We begin our examples with an essential definition.

Definition 1.3.2 A real number a is *rational* if it can be written as a fraction

$$a = \frac{m}{n},$$

for some integers m and n, $n \neq 0$. A number that is not rational is called *irrational*.

■ **Example 1.22** Show that if x is a rational number and y is irrational, the product xy is irrational.

We will present a proof by contradiction. By definition, since x is rational, it can be written as $x = \frac{m}{n}$, where m and n ($n \neq 0$) are integers. As y is irrational, such a representation is not possible.

We want to show that xy is irrational. Assume to the contrary that it is rational. This means, by definition, that there are integers r and s such that $xy = \frac{r}{s}$, $s \neq 0$. So, we have

$$\frac{r}{s} = xy = \frac{m}{n} \cdot y, \text{ or, equivalently } y = \frac{r}{s} \cdot \frac{n}{m} = \frac{rn}{sm}.$$

Since the products rn and sm are integers, we have shown that that y is rational. This contradicts the given fact that y is irrational. We have reached a contradiction. Therefore the product xy is irrational.

■

Exercise 1.27 Show that the sum of a rational and an irrational number is irrational. ∎

> **Exercise 1.28** Show that the ratio of a rational and an irrational number is irrational. ∎

> **Exercise 1.29** Prove (by contradiction) that there are no integers m and n that satisfy $24n + 32m = 4$. ∎

> **Exercise 1.30** If x is rational and xy is irrational, prove that y is irrational. ∎

> **Exercise 1.31** Prove that if $a \in \mathbb{Z}$ and a^3 is even, then a is even. ∎

3.4 Proof by Contradiction of General Statements

In the previous section, we focused on proving or disproving existential and universal statements and presented different techniques for doing so. However, many of the methods remain valid in other settings. In this section we present several classical examples as illustrations for how a proof by contradiction can be constructed for general claims.

■ **Example 1.23** Show that $\sqrt{2}$ is irrational.

We will give a proof by contradiction. We want to prove that

$$p = \sqrt{2} \text{ is irrational}$$

is true. Suppose to the contrary that p is false; that is, suppose that p is rational. By the definition of a rational number, this means that we can write

$$\sqrt{2} = \frac{a}{b}, \tag{1.18}$$

where a and b are integers, $b \neq 0$. Suppose also that the fraction $\frac{a}{b}$ in Equation (1.18) is in lowest terms (call that last claim c).

Now, squaring Equation (1.18) gives

$$2 = \frac{a^2}{b^2}, \quad \text{or, equivalently,} \quad a^2 = 2b^2, \tag{1.19}$$

showing a^2 is even. We know (see Exercise 1.22) that this implies that a is even. That is, we can write $a = 2m$, where m is an integer.

From Equation (1.19), we now have

$$a^2 = (2m)^2 = 4m^2 = 2b^2, \text{ which implies } b^2 = 2m^2. \tag{1.20}$$

This shows that b^2 is even, which implies (by Exercise 1.22 again) that b is even. But a and b both being even means the fraction $\frac{a}{b}$ in Equation (1.18) is *not* in lowest terms (this is claim $\sim c$). Therefore, both c and $\sim c$ must be true, which is impossible. We have reached a contradiction, resulting from our assumption that p is false. Therefore, p must be true, and we have proved that $\sqrt{2}$ is irrational.   ∎

■ **Example 1.24** Prove that there are infinitely many prime numbers.

We will once again use a proof by contradiction assuming that the opposite is true. That is, assume that there is a *finite number* N of primes, and denote those primes by $p_1 < p_2 < \cdots < p_N$; that is, every prime number is on that list[2].

Consider now the number

$$p = p_1 \cdot p_2 \cdots p_N + 1. \tag{1.21}$$

That is, p is the product of all primes on our list plus 1. By the Fundamental Theorem of Arithmetic (see Section 6.6 of the Appendix), this number should have a unique factorization into primes. To construct this factorization, we should first try p_1 as a factor, followed by p_2, etc., until we reach p_N. However, none of these primes divide p (division by each produces a remainder 1). So, p has no prime factors, and, thus, must be itself a prime number. This contradicts the assumption that every prime number is among the listed $p_1, p_2, \ldots, p_n$. Therefore, our assumption was wrong, which proves that there are infinitely many prime numbers.   ∎

Proof by contradiction is also often used to prove uniqueness in existential statements; that is, when we want to prove that there is only one element in a set of interest that satisfies a specific property. In that case, the proof begins with assuming that there are at least two such elements and ends with reaching a contradiction.

■ **Example 1.25** Show that the linear equation

$$ax = b, \quad \text{where } a, b \in \mathbb{R}, \text{ and } a \neq 0,$$

has a unique solution $x \in \mathbb{R}$.

We will prove this by contradiction. Assume there are at least two *different* solutions of the equation: $x_1 \neq x_2$. Since each of them is a solution, we have

$$ax_1 = b$$
$$ax_2 = b.$$

---

[2]This approach follows Euclid's original proof.

Subtracting the second equation from the first leads to

$$ax_1 - ax_2 = 0, \quad \text{or, equivalently,} \quad a(x_1 - x_2) = 0.$$

Since $a \neq 0$, we must have $x_1 - x_2 = 0$, leading to $x_1 = x_2$. We have reached a contradiction with the assumption that there are two different solutions $x_1 \neq x_2$. This proves the solution is unique.   ∎

■ **Example 1.26** This example uses integer division with quotient and a remainder. If not familiar with this material, see Section 6.7 of the Appendix for a quick introduction.

Let a, m be integers, and $m > 0$. The division algorithm states that there are unique integers q and r, with $0 \leq r < m$, for which

$$a = mq + r.$$

We will use proof by contradiction to prove the uniqueness of q and r.

Assume that q and r are not unique. Thus would mean, we can express a in more than one way:

$$a = mq_1 + r_1,$$
$$a = mq_2 + r_2,$$

where $q_1 \neq q_2$ or $r_1 \neq r_2$ with both r_1 and r_2 with values from the set $\{0, 1, \ldots, m-1\}$. Subtracting the two expression, we obtain

$$0 = m(q_1 - q_2) + (r_1 - r_2), \quad \text{or, equivalently,} \quad m(q_1 - q_2) = r_2 - r_1 \quad (1.22)$$

Since $r_1, r_2 \in \{0, 1, \ldots, m-1\}$, their difference satisfies

$$-m < r_2 - r_1 < m.$$

This shows that the quotient $q_1 - q_2$ in Equation (1.22) is 0, so $q_1 = q_2 = q$. Now we get $a = mq + r_1$ and $a = mq + r_2$. Subtracting the second equation from the first gives $r_1 - r_2 = 0$ and, thus, $r_1 = r_2$. This is a contradiction, because we assumed that $q_1 \neq q_2$ or $r_1 \neq r_2$. Therefore, the quotient and the remainder in the division algorithm are unique.   ∎

Exercise 1.32 Prove that $\sqrt[3]{2}$ is irrational. *Hint.* Modify the proof we gave for Example 1.23 and use Exercise 1.31.   ■

Exercise 1.33 Prove that $\sqrt{6}$ is irrational.   ∎

1.3.5 Mathematical Induction

We often need to prove a universal statement $P(n)$ for all integers $n \geq n_0$, where the integer $n_0 \geq 0$ is fixed in advance. For instance, we may want to prove that the following statement holds for all $n \geq n_0$, where $n_0 = 1$:

$$P(n) = \text{``}1 + 2 + \cdots + n = \frac{n(n+1)}{2}.\text{''} \tag{1.23}$$

One might suspect that, indeed, $P(n)$ holds for all $n \geq 1$, since

$$P(1) \text{ states that } 1 = \frac{1(1+1)}{2} = 1,$$

$$P(2) \text{ states that } 1 + 2 = \frac{2(2+1)}{2} = 3,$$

$$P(3) \text{ states that } 1 + 2 + 3 = \frac{3(3+1)}{2} = 6, \text{ and so on,}$$

but how can we *prove* that $P(n)$ holds for *all* $n \geq 1$? We cannot prove it by checking, since there are infinitely many $n \geq 1$.

Another example would be to prove that the statement

$$P(n) = \text{``}7^n - 4^n \text{ is a multiple of 3.''} \tag{1.24}$$

holds for all $n \geq 1$. Again, we may suspect that this is true because

$$P(1) \text{ states that } 7^1 - 4^1 \text{ is a multiple of 3, which is true,}$$
$$\text{since } 7^1 - 4^1 = 3,$$
$$P(2) \text{ states that } 7^2 - 4^2 \text{ is a multiple of 3, which is true,}$$
$$\text{since } 7^2 - 4^3 = 33, \text{ and so on.}$$

However, we cannot prove that $P(n)$ is true for *all* $n \geq 1$ by "checking," as there are infinitely many integers $n \geq 1$.

To construct the proofs of such statements using mathematical induction means that we have to complete the following three steps:

1. Show that $P(n_0)$ is true. This step is sometimes a straightforward verification, but may require more work in some cases. In general, you establish $P(n_0)$ by using any legitimate methods of proof. Establishing that $P(n_0)$ is true is referred to as proving the *base case*.

2. Assume $P(k)$ is true for some $k \geq n_0$. We sometime refer to this as the *inductive assumption*.

3. Use the assumption that $P(k)$ is true to show that $P(k+1)$ is also true.

Notice that steps 2 and 3 are equivalent to proving the following If-Then statement: Prove that if $P(k)$ is true for some integer $k \geq n_0$, then $P(k+1)$ is also true. This is often called the *inductive step* in the proof. Depending on the nature of the problem, the level of difficulty in constructing the proof for the inductive-step may vary considerably. The general idea is that we need to find some connection between $P(k+1)$ and $P(k)$, and use it to show that $P(k+1)$ is true. What is important to keep in mind is that we will need to prove an "If-Then" statement. We can then use any method of proof to do so.

The following theorem justifies why this approach works.

> **Theorem 1.3.1** Steps 1, 2, and 3 above establish that $P(n)$ is true for all $n \geq n_0$.

Proof. We use a proof by contradiction. Assume that steps 1, 2, and 3 have been carried out, but it is not true that $P(n)$ holds for all $n \geq n_0$. This means that for some integers $n \geq n_0$, $P(n)$ is false, so there must be a smallest integer k, for which $P(k)$ is false.

Note that this smallest integer k cannot be n_0, because in step 1 we have verified that the base case $P(n_0)$ is true.

Note also that since k is the smallest integer for which $P(n)$ is false, $P(k-1)$ must be true. But now, by steps 2 and 3, we know that if $P(k-1)$ is true, then $P(k)$ must be true too (because $k = (k-1)+1$). We have reached a contradiction with $P(k)$ being false. Thus, following steps 1, 2, and 3 establishes that $P(n)$ is true for all $n \geq n_0$. ∎

■ **Example 1.27** Prove that the statement $P(n)$ from Equation (1.23) is true for all $n \geq 1$.

In this case $n_0 = 1$. We follow the required steps:

1. We have to show that $P(1)$ is true. This is indeed the case because for $n = 1$, the sum on the left has only one term, so $P(1) = 1$. When $n = 1$, we also have (as we already checked above)

$$\frac{n(n+1)}{2} = \frac{1(1+1)}{2} = 1.$$

2. Assume $P(k)$ is true for some $k \geq n_0$. That is, assume that the following holds:

$$P(k) = 1 + 2 + \cdots + k = \frac{k(k+1)}{2}. \tag{1.25}$$

Note that we don't have to do anything more here; this step simply postulates that $P(k)$ is true.

3. We now have to prove that if $P(k)$ is true, then $P(k+1)$ is also true. In other words, we need to use Equation (1.25) to prove that the sum $P(k+1)$ can be written as

$$P(k+1) = \frac{(k+1)[(k+1)+1]}{2}. \tag{1.26}$$

We now have

$$P(k+1) = \underbrace{1+2+\cdots+k}_{\frac{k(k+1)}{2}}+(k+1)$$

$$\text{(Using } 1+2+\cdots+k = \frac{k(k+1)}{2}, \text{ since } P(k) \text{ is true.)}$$

$$= \frac{k(k+1)}{2} + (k+1)$$

$$= \frac{k(k+1)+2(k+1)}{2}$$

$$= \frac{(k+1)(k+2)}{2}$$

$$= \frac{(k+1)[(k+1)+1]}{2},$$

which establishes Equation (1.26). Thus, $P(k+1)$ is true. Therefore we have proved, by mathematical induction, that

$$1+2+\cdots+n = \frac{n(n+1)}{2}, \text{ for all integers } n \geq 1.$$

∎

■ **Example 1.28** Prove that $7^n - 4^n$ is a multiple of 3 for all $n \geq 1$.

In this case, we have $n_0 = 1$, and we follow the required steps:

1. Check that the base case $P(1)$ is true. This is straightforward (as we already checked above), since for $n = 1$, we have

$$7^n - 4^n = 7^1 - 4^1 = 3, \text{ which is a multiple of 3.}$$

2. Assume that $P(k)$ is true for some $k \geq 1$. That is, assume that $7^k - 4^k$ is a multiple of 3, which means that $7^k - 4^k = 3m$, for some integer m.

3. We now have to show that $P(k)$ being true implies that $P(k+1)$ is also true.

To prove $P(k+1)$, we have to show that $7^{k+1} - 4^{k+1}$ is a multiple of 3. We write

$$7^{k+1} - 4^{k+1} = 7(7^k) - 4(4^k) = 3(7^k) + 4(7^k) - 4(3^k) = 3(7^k) - 4(7^k - 3^k).$$

By the inductive assumption, we have that $7^k - 4^k = 3m$, so

$$7^{k+1} - 4^{k+1} = 3(7^k) - 4(7^k - 3^k) = 3(7^k) - 4(3m) = 3(7^k - 4m).$$

We see that $7^{k+1} - 4^{k+1}$ is a multiple of 3, establishing that $P(k+1)$ is true.

Thus, by mathematical induction, we have proved that $7^n - 3^n$ is a multiple of 3 for all $n \geq 1$.

∎

Exercise 1.34 Prove by induction that

$$1^2 + 2^2 + 3^2 + \cdots + n^2 = \frac{n(n+1)(2n+1)}{6}, \text{ for all } n \geq 1.$$

∎

Exercise 1.35 Prove by induction that for every positive integer n

$$\frac{1}{1 \cdot 2} + \frac{1}{2 \cdot 3} + \frac{1}{3 \cdot 4} + \cdots + \frac{1}{n(n+1)} = \frac{n}{n+1}.$$

∎

Exercise 1.36 Use mathematical induction to show that for every non-negative integer n

$$3 + (3+5) + (3+5 \cdot 2) + (3+5 \cdot 3) + \cdots + (3+5 \cdot n) = \frac{(n+1)(5n+6)}{2}.$$

∎

Exercise 1.37 Prove, by mathematical induction, that for any $n \geq 1$

$$1 + 3 + 5 + \cdots + (2n-1) = n^2.$$

∎

Exercise 1.38 Use mathematical induction to prove that $n^3 + 5n + 6$ is a multiple of 3 for any integer $n \geq 1$.

1.3.6 Proving If-And-Only-If Statements

Recall from Section 1.2.6 that

$$p \Leftrightarrow q = (p \Rightarrow q) \wedge (q \Rightarrow p). \tag{1.27}$$

This means that for $p \Leftrightarrow q$ to be true, both $p \Rightarrow q$ and $q \Rightarrow p$ must be true. Thus, proofs of *if-and-only-if* statements can be constructed by providing two proofs.

Proof 1. Prove the *if-then* statement $p \Rightarrow q$;

Proof 2. Prove the *if-then* statement $q \Rightarrow p$.

The two proofs constitute the proof of the *if-and-only-if* statement $p \Leftrightarrow q$.

■ **Example 1.29** Prove that if x, y are integers, the sum $x + y$ is an odd number if and only if exactly one of x and y is even, and the other is odd.

This is an *if-and-only-if* statement and we will have to prove two *if-then* statements. That is, we need to prove

1. If $x + y$ is odd, then exactly one of x and y is even (so, the other one is odd), and

2. If exactly one of x and y is odd, then the sum $x + y$ is odd.

Proof 1. We will give a proof by contradiction. Let $x + y$ be odd, and assume that it is not true that exactly one of x and y is even. Then x and y will be both even or both odd. We will examine each of these cases separately.

Let x, and y be both even; that is, $x = 2k$, for $k \in \mathbb{Z}$ and $y = 2m$, for $m \in \mathbb{Z}$. Then $x + y = 2k + 2m = 2(k + m)$, so the sum is even. This is a contradiction, since we know $x + y$ is odd.

Now let x, and y be both odd; that is, $x = 2k + 1$, for $k \in \mathbb{Z}$ and $y = 2m + 1$, for $m \in \mathbb{Z}$. Then $x + y = (2k + 1) + (2m + 1) = 2(k + m + 1)$, so the sum in this case is also even. This is a contradiction, since we know $x + y$ is odd.

Proof 2. We need to prove that if exactly one of x and y is even, then $x + y$ is odd. We will use a direct proof. Let's assume, that x is even and y is odd; that is $x = 2k$, for $k \in \mathbb{Z}$ and $y = 2m + 1$, for $m \in \mathbb{Z}$. Then, $x + y = 2k + (2m + 1) = 2(k + m) + 1 = 2n + 1$, where $n = k + m \in \mathbb{Z}$. This proves that $x + y$ is odd. Similarly, we can see that $x + y$ is odd when x is odd and y is even.

Proofs 1 and 2 show that $x + y$ is odd if and only if exactly one of x and y is even.   ■

Note that we used different methods of proof for Proof 1 and Proof 2 above. This is not unusual. In principle, we can construct Proof 1 and Proof 2 by using any legitimate method of proof, like those described in previous sections.

Exercise 1.39 Prove that an integer a is odd if and only if a^2 is odd. ∎

Exercise 1.40 Prove that and integer b is even if and only if b^2 is even. ∎

Exercise 1.41 Prove that the product of two integers is odd if and only if they are both odd. ∎

Exercise 1.42 Let x and y be integers, neither of which is a multiple of 3. Prove that $x + y$ is a multiple of 3 if and only if the integer reminders of x and y from division by 3 add up to 3. *Hint.* Use the division algorithm (Theorem 6.7.1 in the Appendix). ∎

.4 Proofs Involving Sets

We often need to prove that two sets that have been defined in the context of a specific question are equal or that one set is a subset of another. You will recall from Section 1.1.1 that two sets are equal when they contain the same elements, and Definition 1.1.3 formalizes our understanding of set inclusion. When one deals with finite sets with small number of elements, it is often possible to verify set equality or set inclusion by simple inspection. For larger or infinite sets, and when sets are defined as

$$A = \{x \mid x \text{ satisfies a certain property } P\},$$

verifying by inspection is often difficult or impossible.

.1 Proving a Set is a Subset of Another Set

Definition 1.1.3 tells us that $A \subseteq B$ if, for any $x \in A$, we can show $x \in A \Rightarrow x \in B$.

Thus, when we choose to use direct proofs, a proof for $A \subseteq B$ always begins with "Let $x \in A$." and should end with "Therefore $x \in B$." In-between, we will use definitions, the specific properties of the sets A and B, and logic to supply the chain of logical arguments providing the proof.

■ **Example 1.30** Let

$$A = \{x \in \mathbb{Z} \mid x = 6m, \text{ where } m \in \mathbb{Z}\}, \text{ and } B = \{x \in \mathbb{Z} \mid x = 2n, \text{ where } n \in \mathbb{Z}\}.$$

Prove that $A \subseteq B$.

Note that both sets are infinite, so we cannot determine the inclusion by simple inspection. We will use a direct proof to show $A \subseteq B$.

Let $x \in A$. This means, there is an integer m such that $x = 6m$. We can now write $x = 2(3m) = 2n$, where $n = 3m$ is an integer. Thus, $x \in B$. Therefore, we have proved that $A \subseteq B$.

In this case, we can also show that the inclusion is strict, that is $A \subset B$. To do this, we need to find an element of B that is not in A. One such element x is $x = 10$ (and there are many others). We can see that $x \in B$, since $x = 2 \cdot 5$, but $x \notin A$, since 10 is not a multiple of 6. This proves that $A \subset B$.   ■

■ **Example 1.31** Let A and B be arbitrary sets with $A \cap B \neq \varnothing$. Prove that $A \cap B \subseteq B$ and $A \cap B \subseteq A$.

We'll again do a direct proof. Let $x \in A \cap B$. By the definition of set intersection (see Section 1.1.2), this means we have $x \in A$ and $x \in B$, so we also have $x \in B$. Therefore $A \cap B \subseteq B$. The same argument can be used to show $A \cap B \subseteq A$.

    ■

We present more examples of this nature in Section 1.4.4 in the context of sets that are obtained as images or preimages of functions.

Exercise 1.43 Let

$$A = \{x \in \mathbb{Z} \mid x = 8m, \text{ where } m \in \mathbb{Z}\}, \text{ and } B = \{x \in \mathbb{Z} \mid x = 4n, \text{ where } n \in \mathbb{Z}\}.$$

Prove that $A \subseteq B$. Next, show that the inclusion is strict by finding an element of B that is not in A.   ▨

Exercise 1.44 Let

$$A = \{4^n \mid n \in \mathbb{Z}\}, \text{ and } B = \{2^n \mid n \in \mathbb{Z}\}.$$

Prove that $A \subseteq B$. Next, show that the inclusion is strict by finding an element of B that is not in A.   ▨

Exercise 1.45 Let A, B, C be sets. If $B \subseteq C$, prove that $A \times B \subseteq A \times C$. ∎

4.2 Proving That Two Sets Are Equal

To prove that two sets A and B are equal, we usually show that each of the sets is a subset of the other. So, to show that $A = B$, we will need to show that $A \subseteq B$ and that $B \subseteq A$ using the technique we outlined above.

∎ **Example 1.32** Prove that for any sets A, B, and C,

$$(A \cup B) \cap C = (A \cap C) \cup (B \cap C).$$

See Figure 1.4 where we depicted the Venn diagram corresponding to these sets. We will restate again that the Venn diagram does not provide a proof for the two sets being equal, but it is a great tool to suggest identities between sets that can then be proved by the method we discuss here.

Step 1. We will prove

$$(A \cup B) \cap C \subset (A \cap C) \cup (B \cap C). \tag{1.28}$$

Let $x \in (A \cup B) \cap C$. By definition of intersection of sets, this means that $x \in A \cup B$ and $x \in C$. Having $x \in A \cup B$, by definition, means that $x \in A$ or $x \in B$. Let's assume $x \in A$ (the argument for the case $x \in B$ can be carried out similarly). This means $x \in A \cap C$. Now, by the definition of union of sets, since $x \in A \cap C$, we have $x \in (A \cap C) \cup (B \cap C)$. This proves the inclusion from Equation (1.28).

Step 2. We will now prove

$$(A \cap C) \cup (B \cap C) \subset (A \cup B) \cap C. \tag{1.29}$$

Let $x \in (A \cap C) \cup (B \cap C)$. By the definition of union, this means that $x \in A \cap C$ or $x \in B \cap C$. Let's assume $x \in A \cap C$ (the argument for $x \in B \cap C$ can be carried out in the exact same way). This means, by the definition of set intersection, that $x \in A$ and $x \in C$. Next, since $x \in A$, by the definition of set union, $x \in A \cup B$. Now, since $x \in A \cup B$ and $x \in C$, we have shown that $x \in (A \cup B) \cap C$. This proves the inclusion from Equation (1.29). Combining steps 1 and 2, proves the two sets are equal.   ∎

In the last two examples, when we got to a point where we showed an element x is in the union of two sets, we chose to continue under the assumption that it belongs to one of the sets, and noted that assuming it is in the other, can be handled in the exact same way. For such situations, we often use the phrase "Without loss of generality, we can assume..." We use this language in our next example.

■ **Example 1.33** For any two sets, A and B, prove that

$$(A \cap B)^c = A^c \cup B^c. \tag{1.30}$$

Let $x \in (A \cap B)^c$. This means $x \notin (A \cap B)$ so it cannot be in both sets. Without loss of generality, let's assume $x \notin A$, which is the same (by the definition of a set complement) as $x \in A^c$. Since $x \in A^c$, by definition of set union, $x \in A^c \cup B^c$. Therefore, we have proved that

$$(A \cap B)^c \subset A^c \cup B^c. \tag{1.31}$$

Now let $x \in A^c \cup B^c$. Without loss of generality, we can assume that $x \in B^c$, which is the same as $x \notin B$. But if x is not in the set B, then $x \notin A \cap B$. Therefore $x \in (A \cap B)^c$. This proves

$$A^c \cup B^c \subset (A \cap B)^c. \tag{1.32}$$

The inclusions from Equations (1.31) and (1.32) prove the claim from Equation (1.30), known as one of the *DeMorgan's laws for sets*. We leave the proof of the other DeMorgan's law for sets as an exercise (see Exercise 1.46).

■

Exercise 1.46 (*DeMorgan's law for sets*) Prove that for any two sets A and B,

$$(A \cup B)^c = A^c \cap B^c$$

■

Exercise 1.47 Prove that for any sets A, B, and C, the following distributive properties hold:

1. $A \cap (B \cup C) = (A \cap B) \cup (A \cap C)$;

2. $A \cup (B \cap C) = (A \cup B) \cap (A \cup C)$;

3. $(A \cup B) \cap C = (A \cap C) \cup (B \cap C)$.

■

> **Exercise 1.48** Let A, B, C be sets. Prove that
>
> $$(A \times B) \cap (A \times C) = A \times (B \cap C).$$
>
> ∎

> **Exercise 1.49** Let A, B, C be sets. Prove that
>
> $$(A \cap B) - C = (A - C) \cap (B - C).$$
>
> ∎

.3 Proving that Two Sets Are Disjoint

Recall that two sets A and B are disjoint, if $A \cap B = \varnothing$. To prove that, we often use a proof by contradiction. We assume that the intersection is not empty and this leads us to a contradiction. Such proofs begin with the sentence "Assume to the contrary that $A \cap B \neq \varnothing$ and $x \in A \cap B$."

∎ **Example 1.34** Consider the sets

$$A = \{x \mid x = 12m + 4, \ m \in \mathbb{Z}\} \quad \text{and} \quad B = \{x \mid x = 24k, \ k \in \mathbb{Z}\}.$$

Prove that $A \cap B = \varnothing$.

Assume to the contrary that $A \cap B \neq \varnothing$ and $x \in A \cap B$. This means $x \in A$ and $x \in B$. Since $x \in A$, there exists an integer m, such that

$$x = 12m + 4.$$

Since $x \in B$, there exists an integer k, such that

$$x = 24k.$$

So, for these m and k, we must have

$$12m + 4 = 24k, \text{ or, equivalently } 3m + 1 = 6k = 3(2k). \tag{1.33}$$

But Equation (1.33) is impossible, since the integer on the right is divisible by 3 and the number on the left gives a remainder 1 when divided by 3. We have reached a contradiction, which proves that $A \cap B = \varnothing$.   ∎

∎ **Example 1.35** Let A and B be non-empty sets. Prove that the sets $A - B$ and $B - A$ are disjoint.

As we have already said a number of times, we could look at the Venn diagram (see Figure 1.1) for insight, but it does not provide a proof. The rigorous proof goes as follows:

Assume to the contrary that the two sets are not disjoint and there is an $x \in (A - B) \cap (B - A)$. This means $x \in A - B$ and $x \in B - A$.

By the definition of set difference, $x \in A - B$ means that

$$x \in A \text{ and } x \notin B. \tag{1.34}$$

On the other hand, $x \in B - A$ means that

$$x \in B \text{ and } x \notin A. \tag{1.35}$$

The statements in Equations (1.34) and (1.35) contradict one another, so they cannot be true at the same time. Therefore $A - B$ and $B - A$ are disjoint. ∎

Exercise 1.50 Given two sets A and B, prove that $A \cap (B - A) = \varnothing$.

Exercise 1.51 Prove that for any set A, $A \cap \varnothing = \varnothing$.

Exercise 1.52 Let $A = \{a \in \mathbb{Z} \mid a = 3k + 2, \ k \in \mathbb{Z}\}$ and $B = \{b \in \mathbb{Z} \mid b = 9m + 7, \ m \in \mathbb{Z}\}$. Prove that $A \cap B = \varnothing$.

Exercise 1.53 Let $A = \{a \in \mathbb{Z} \mid a = 5k + 1, \ k \in \mathbb{Z}\}$ and $B = \{b \in \mathbb{Z} \mid b = 15m + 4, \ m \in \mathbb{Z}\}$. Prove that $A \cap B = \varnothing$.

Exercise 1.54 Let $A = \{2^n \mid n \in \mathbb{Z}\}$ and $B = \{3^m \mid m \in \mathbb{Z}\}$. Prove that $A \cap B = \varnothing$.

1.4.4 Proofs Involving Inverse Images of Functions

This section is optional. It provides a series of examples where the goal is to prove inclusion or equality for sets arising as images of functions and their inverses. The material is somewhat more technical than what we have presented so far, but we have included it here for two reasons: (1) It provides additional authentic examples where the methods of proofs discussed in this section can be applied at a more advanced level and (2) This topic is often short-changed in real analysis courses due to a lack of time. You may choose to go through the proofs now or return to this section at a later point when the properties discussed in the section are needed for constructing more advanced proofs.

A refresher on the terminology and definitions regarding functions, as well as images and inverse images of sets for a given function, can be found in Section 6.8 of the Appendix. We recommend reviewing this section before proceeding with what follows below.

Proposition 1.4.1 Let $f : X \to Y$ be a function, and B_1, B_2 be subsets of Y. Show that

$$f^{-1}(B_1 \cup B_2) = f^{-1}(B_1) \cup f^{-1}(B_2). \tag{1.36}$$

Proof. In Section 1.4.2 we discussed that to prove that two sets A and B are the same, we need to show that any $x \in A$ is also in B and that any $x \in B$ is also in A. We can also express this as having to show that $x \in A$ if and only if $x \in B$.

Now let $x \in f^{-1}(B_1 \cup B_2)$. From the definitions of image and pre-image of f, we have:

$$
\begin{array}{ll}
x \in f^{-1}(B_1 \cup B_2), & \text{if and only if} \\
f(x) \in B_1 \cup B_2, & \text{if and only if} \\
f(x) \in B_1 \text{ or } f(x) \in B_2, & \text{if and only if} \\
x \in f^{-1}(B_1) \text{ or } f^{-1}(B_2), & \text{if and only if} \\
x \in f^{-1}(B_1) \cup f^{-1}(B_2). &
\end{array}
$$

This proves

$$f^{-1}(B_1 \cup B_2) = f^{-1}(B_1) \cup f^{-1}(B_2).$$

∎

Proposition 1.4.2 Let $f : X \to Y$ be a function, and A_1, A_2 be subsets of X. Show that

$$f(A_1 \cap A_2) \subseteq f(A_1) \cap f(A_2). \tag{1.37}$$

Proof. Let $y \in f(A_1 \cap A_2)$. From Definition 6.12.1 in the Appendix, this means that there is an $x \in A_1 \cap A_2$ with $y = f(x)$.

Since $x \in A_1$, we know $y = f(x) \in f(A_1)$. But we also have $x \in A_2$, which means $y = f(x) \in f(A_2)$. Thus, $y \in f(A_1) \cap f(A_2)$. This proves the inclusion

$$f(A_1 \cap A_2) \subseteq f(A_1) \cap f(A_2).$$

∎

We should note that the sets $f(A_1 \cap A_2)$ *may be a proper subset* of $f(A_1) \cap f(A_2)$ for some functions f. As our next example shows, there may be a $y \in f(A_1) \cap f(A_2)$ that is *not* in $f(A_1 \cap A_2)$.

■ **Example 1.36** Consider the function $f(x) = x^2$, defined over the real numbers. Let $A_1 = \{-2\}$ and $A_2 = \{2\}$. Then $f(A_1) = f(A_2) = \{4\}$; that is, $f(A_1) \cap f(A_2) = \{4\}$. But $A_1 \cap A_2 = \varnothing$, which implies $f(A_1 \cap A_2) = \varnothing$. Thus, we have found a function for which

$$f(A_1 \cap A_2) \neq f(A_1) \cap f(A_2).$$

Note that the same argument could be used to show that if a function f is not one-to-one (see Definition 6.9.2), $f(A_1 \cap A_2) \neq f(A_1) \cap f(A_2)$ (see Exercise 1.55).   ■

Proposition 1.4.3 Let $f : X \to Y$ be a function, and $B \subseteq Y$. Show that

$$f^{-1}(B^c) = [f^{-1}(B)]^c. \tag{1.38}$$

Proof. Using Definition 6.12.2 in the Appendix, we again give a sequence of if-and-only-if statements that establishes the claim. We have

$$\begin{aligned}
x &\in f^{-1}(B^c), &&\text{if and only if} \\
f(x) &\in B^c, &&\text{if and only if} \\
f(x) &\notin B, &&\text{if and only if} \\
x &\notin f^{-1}(B), &&\text{if and only if} \\
x &\in [f^{-1}(B)]^c,
\end{aligned}$$

which proves that $f^{-1}(B^c) = [f^{-1}(B)]^c$.   ■

Proposition 1.4.4 Let $f : X \to Y$ be a function. Prove that

$$A \subseteq f^{-1}(f(A)) \tag{1.39}$$

Proof. Let $x \in A$. Then $y = f(x) \in f(A)$ and, by Definition 6.12.2, $x \in f^{-1}(f(A))$. Thus, $A \subseteq f^{-1}(f(A))$.   ■

Note that the inclusion in Equation (1.39) may be strict for some functions. So, in general, the two sets may be different. Our next example shows that.

■ **Example 1.37** Let's look again at $f : \mathbb{R} \to \mathbb{R}$, $f(x) = x^2$. Let $A = \{1\}$. Then $f(A) = \{1\}$, while $f^{-1}(f(A)) = f^{-1}(\{1\}) = \{-1, 1\} \neq A$.   ■

Note that a similar example may be given for any function that is not one-to-one. When f *is* one-to-one, $A = f^{-1}(f(A))$ for any $A \subseteq X$. We leave the proof as an exercise (see Exercise 1.58).

Proposition 1.4.5 Let $f : X \to Y$ be a function, and $B \subseteq Y$. Show that

$$f(f^{-1}(B)) \subseteq B. \tag{1.40}$$

Proof. Let $y \in f(f^{-1}(B))$. We'll show that $y \in B$. Since $y \in f(f^{-1}(B))$, there is an $x \in f^{-1}(B)$, such that $y = f(x)$. Since $x \in f^{-1}(B)$, we have $f(x) \in B$, by Definition 6.12.2. But $f(x) = y$, and therefore $y \in B$.   ∎

Just as we have noted in several other occasions already, the inclusion in Equation (1.40) may be strict for some functions. Our next example shows that, in general, the two sets may be different.

■ **Example 1.38** Let $f : X \to Y$ be the function $f(x) = x^2$, where $X = Y = \mathbb{R}$. Now let $B = (-3, \infty) \subset Y = \mathbb{R}$. Then $f^{-1}(B) = \{x \in \mathbb{R} \mid x^2 \in (-3, \infty)\} = (-\infty, \infty)$. But $f(f^{-1}(B)) = \{y \in \mathbb{R} \mid y = f(x), \text{ for some } x \in (-\infty, \infty)\} = [0, \infty) \neq B = (-3, \infty)$.   ■

You may have noticed that this example works because the function $f = x^2$, $f : \mathbb{R} \to \mathbb{R}$, is not onto (see Definition 6.9.3). When a function $f : X \to Y$ *is* onto, we will always have $f(f^{-1}(B)) = B$. We leave the proof as an exercise (see Exercise 1.59).

Theorem 1.4.6 Let the function $f : X \to Y$ be one-to-one and onto (that is, f is a bijection). Then for any $A \subseteq X$ and any $B \subseteq Y$,

$$A = f^{-1}(f(A)) \text{ and } B = f(f^{-1}(B)).$$

Proof. We leave the proof as an exercise (see Exercises 1.58 and 1.59).   ∎

Exercise 1.55 Show that if a function $f : X \to Y$ is *not* one-to-one, we can find subsets A_1, A_2 of X for which $f(A_1 \cap A_2) \neq f(A_1) \cap f(A_2)$.   ■

Exercise 1.56 Give more examples of functions $f : X \to Y$ and $A \subset X$, where $A \neq f^{-1}(f(A))$.   ■

Exercise 1.57 Give more examples of functions $f : X \to Y$ and $B \subset Y$, where $B \neq f(f^{-1}(B))$.   ■

> **Exercise 1.58** Let $f : X \to Y$ be a one-to-one function. Then for any $A \subseteq X$, we have the equality $A = f^{-1}(f(A))$. ∎

> **Exercise 1.59** Show that if the $f : X \to Y$ is onto, then $f(f^{-1}(B)) = B$. ∎

1.5   Summary and What to Expect Next

In this chapter, we gave you some important building blocks that constitute the foundation of theoretical mathematics. As you will see in later chapters, sets and their properties will be used to define important abstract structures, while the highlights from mathematical logic we presented will help us construct rigorous mathematical proofs. The vocabulary, understanding existential and universal statements, as well as understanding how to negate compound logical statement and how to best use the main methods of mathematical proofs, form the basic toolkit of abstract mathematics. Although often in the background, facility in their use is invaluable.

What we have presented here is often called naive set theory and we have only provided the basic essentials from mathematical logic. These are, however, two broad mathematical disciplines, from which, if you choose to study them in more depth, you can learn about axiomatic approaches to set theory, formal number theory, recursion, the notion of computability, and the so-called model theory. For what follows, we will use the material from this chapter mainly as a tool in the study of other mathematical disciplines. We hope, however, that you have found the content of enough interest to want to pursue some of its more advanced topics on your own.

1.6   Suggested Further Reading

We recommend the following textbooks for further reading. There are also many others. They are standard texts used for undergraduate Discrete Mathematics and Introduction to Proofs courses.

Bibliography

[1] Hammack, R. *Book of Proof*, 3rd Ed. eBook (Creative Commons Licensed), 2018.

[2] Hunter, David J. *Essentials of discrete mathematics*. Jones & Bartlett Learning, 2021.

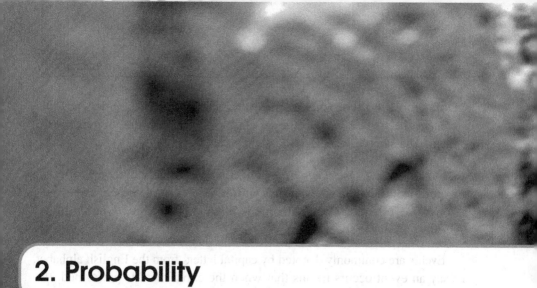

2. Probability

Probability theory has its roots in gambling and the desire of gamblers to have an accurate estimate of their chances of winning. The main assumption in such cases is that the game can be built from equally likely outcomes such as flipping a coin, rolling a die, drawing a card from a standard deck, or using a spinner with a number of circular segments of equal angles. Such problems form the field of Elementary Probability. Nowadays, probability is widely used in nearly every aspect of life, and not all problems can be reduced to equally likely outcomes. The unified theory that can handle either case requires a rigorous axiomatic approach.

In this chapter, we present the axioms of probability along with some applications. Our main focus however is on solving elementary probability problems where it is essential to be able to count the number of outcomes that "generate" a complex event. In that context, we discuss counting arguments, independent events, discrete random variables, conditional probability, and computation of expected value.

1 Sample Spaces and Events

The setting of probability is that we have an experiment whose outcome depends on chance. As in any other field of mathematics, probability uses a special vocabulary that will be pervasive throughout this section.

DOI: 10.1201/9781032623849-2

Definition 2.1.1 For an experiment, the outcome of which depends on chance, the *sample space* is the set of all possible outcomes to the experiment. The sample space is typically denoted by S or Ω.

Definition 2.1.2 An *event* is a collection of possible outcomes for an experiment. That is, an event is a subset of the sample space. An event consisting of exactly one outcome is often called a *sample point* or an *elementary event*.

Events are commonly denoted by capital letters from the English alphabet. To say an event occurs means that when the experiment is performed, the outcome is one of the elements of the set.

■ **Example 2.1** We roll a six-sided die. What outcome we will get is our experiment. The sample space is $\Omega = \{1, 2, 3, 4, 5, 6\}$. The elementary events are the one element subsets of Ω: $\{1\}$, $\{2\}$, $\{3\}$, $\{4\}$, $\{5\}$, $\{6\}$. Let's denote by A the event "An even number is rolled." Then $A = \{2, 4, 6\}$. If we consider the subset $B = \{3, 6\}$, the event B can be described in words as "We roll a number divisible by 3."

Let's say, we rolled a 3. With the above notation, the event B and the elementary event $\{3\}$ occurred while the event A didn't.   ■

■ **Example 2.2** We perform the following experiment: flip a coin three times and record the outcome as a sequence. Let's say we get THH, which means we got tails on the first flip, heads on the second, and heads on the third flip. This is just one of several possible outcomes from the experiment. To get the sample space Ω, we will need to list *all* possible sequences of length three that can be constructed from the letters H and T. Thus

$$\Omega = \{HHH, HHT, HTH, HTT, THH, THT, TTH, TTT\}.$$

If E is the event "We obtain at least two tails," this event is represented by the subset $E = \{HTT, TTH, THT, TTT\}$. The event $F = \{HHH, HHT, HTH, HTT, THH, THT, TTH\}$ can be described in words as "We obtain one or more heads."

With our outcome THH, the event E did not occur, while the event F did.   ■

Definition 2.1.3 Two events A and B are *disjoint* if they do not have an outcome in common; that is, if $A \cap B = \varnothing$. A number of events are *pairwise disjoint* when no two of the events have an outcome in common.

Since events are subsets of the experiment's sample space, you may have already noticed that we use the same notation as for sets. We can also create new sets using operations on sets. For the sets from Example 2.2, we can consider the new events $G = E \cap F = \{HTT, TTH, THT\}$ – "Two tails and one heads are obtained," and $E \cup F = \{HHH, HHT, HTH, HTT, THH, THT, TTH, TTT\} = \Omega$. We also use the subset notation and write, e.g., $G \subset E$, and we use $A^c = \Omega - A$ to denote the complement of A. For Example 2.2, $F^c = \{TTT\}$.

2.2 The Probability Function

In many of the examples that we will consider in this chapter the sample space will be much larger than those above, and, in what follows, we will develop methods for finding the number of elements in a sample space without listing all of them. We also discuss how to interpret and determine the likelihood for a certain event to occur.

Definition 2.2.1 A function P, that assigns to each event $A \subset \Omega$ a value in the interval $[0, 1]$ is called a *probability* if it satisfies the following axioms.
1. For any $A \subseteq \Omega$, $0 \leq P(A) \leq 1$;
2. $P(\Omega) = 1$;
3. For a sequence of pairwise disjoint events $A_1, A_2, A_3, \ldots,$

$$P\left(\bigcup_{i=1}^{\infty} A_n\right) = \sum_{i=1}^{\infty} P(A_n).$$

As a special case, when A and B are disjoint events, $P(A \cup B) = P(A) + P(B)$.

We first prove some consequences from the axioms.

Theorem 2.2.1 Let A and B be events in a sample space Ω. Then the following properties hold:
1. If $A \subseteq B$, then $P(A) \leq P(B)$;
2. $P(B) = P(A \cap B) + P(B - A)$;
3. $P(A^c) = 1 - P(A)$;
4. $P(\emptyset) = 0$;
5. $P(A \cup B) = P(A) + P(B) - P(A \cap B)$.

Proof.   1. When $A \subseteq B$, $B - A$ and A are disjoint (see Figure 2.1). From Axiom 3, $P(B) = P(A) + P(B - A) \geq P(A)$. The inequality is true because $P(B - A) \geq 0$, according to the first axiom.
2. We can write $B = (B - A) \cup (A \cap B)$, and $B - A$ and $A \cap B$ are disjoint (see Figure 2.2). So, by axiom 1, $P(B) = P(A \cap B) + P(B - A)$.

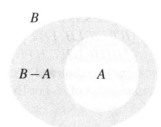

Figure 2.1: When $A \subseteq B$, $B = (B - A) \cup A$, and $B - A$ and A are disjoint.

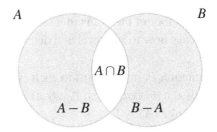

Figure 2.2: Sets $B - A$ and $A \cap B$ are disjoint.

3. Since $A \cup A^c = \Omega$, and A and A^c are disjoint (see Figure 2.3), axioms 1 and 3 imply

$$1 = P(\Omega) = P(A) + P(A^c), \quad \text{so } P(A^c) = 1 - P(A).$$

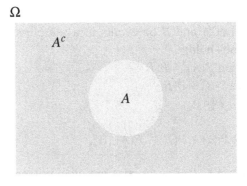

Figure 2.3: The sets A and A^c are disjoint, and $A \cup A^c = \Omega$.

4. From the property above, if A is an event, $A^c = 1 - P(A)$. When $A = \emptyset$, $A^c = \Omega$. From axiom 2, $P(\Omega) = 1$, thus $P(\emptyset) = 1 - P(\Omega) =$

$1 - 1 = 0$. The event represented by the empty set $\varnothing$ is called the *impossible event*.

5. Let's first give the intuition behind this. We know from axiom 3 that if A and B are disjoint, $P(A \cup B) = P(A) + P(B)$. However, as the sets A and B may overlap as in Figure 2.2, using again $P(A \cup B) = P(A) + P(B)$ would mean we have counted $A \cap B$ twice. Thus, we need to subtract $P(A \cap B)$ to make things right.

More formally, notice that

$$A \cup B = (A - B) \cup (B - A) \cup (A \cap B),$$

and the sets are disjoint (see Figure 2.2). Thus, by axiom 3,

$$P(A \cup B) = P(A - B) + P(B - A) + P(A \cap B).$$

Now, since $P(A) = P(A - B) + P(A \cap B)$, and since $P(B) = P(B - A) + P(A \cap B)$, we obtain

$$\begin{aligned}
P(A \cup B) &= P(A - B) + P(B - A) + P(A \cap B) \\
&= [P(A - B) + P(A \cap B)] + [P(B - A) + P(A \cap B)] \\
&\quad - P(A \cap B) \\
&= P(A) + P(B) - P(A \cap B).
\end{aligned}$$

∎

2.3 Equally Likely Outcomes

A common occurrence is that the sample space consists of sample points that all have the same probability. We now discuss the consequences of this.

Consider again Example 2.1. The outcomes are $1, 2, 3, 4, 5, 6$. For a fair die, all outcomes (elementary events) should be equally likely, so we assign

$$P(i) = \frac{1}{6}, \text{ for } i = 1, 2, 3, 4, 5, 6.$$

Since any two elementary events are disjoint, the probability of any event is the sum of the probabilities of the elementary events that it is comprised of. For example,

$$P(\{1, 2, 6\}) = P(\{1\}) + P(\{2\}) + P(\{6\}) = \frac{1}{6} + \frac{1}{6} + \frac{1}{6} = \frac{3}{6} = \frac{1}{2}.$$

This example shows that if the number of possible outcomes from an experiment is finite and all outcomes are equally likely, then each elementary event has probability $\frac{1}{|\Omega|}$, where $|\Omega|$ denotes the number of elements in the sample space.

This leads to the following definition.

Definition 2.3.1 Suppose that a sample space Ω has n elements and each outcome is equally likely. If an event A has k elements, then the probability of the event A is defined to be

$$P(A) = \frac{k}{n} = \frac{|A|}{|\Omega|}.$$

We need to stress that not all experiments have the property that all elementary events are equally likely. When they do, however, the probability of each elementary event (data point) is $\frac{1}{|\Omega|}$. In what follows, we will focus exclusively on this case.

Definition 2.3.1 tells us that to find the probability of an event, we need to count how many elements the event and the sample space have. In the examples we considered so far, we were able to list all outcomes and then count them. When the sample space is large, this method is not practical. In what follows, we will develop some techniques that will allow us to determine the number of elements in a set without having to list them. One such approach is to think that our "experiment" occurs in stages, which is a common occurrence.

To begin, we state a very basic counting principle.

Theorem 2.3.1 **Multiplication Principle**. Suppose two tasks can be performed in a sequence. If Task 1 can be performed in m ways and Task 2 can be performed in n ways, the sequence of the tasks can be performed in mn ways.

■ **Example 2.3** You want to fix a quick dinner and you have frozen pizza crusts. You also have two types of pizza sauce – classic marinara and BBQ, and you have pepperoni, anchovies, and ham as toppings. How many different one-topping pizzas can you create?

We can break the process of preparing a pizza into two tasks: Task 1 – select a pizza sauce, and Task 2 – select a topping. A sauce can be selected in two ways. Once a sauce is selected, there are three possible ways for selecting a topping, giving $(2)(3) = 6$ different pizza combinations (see Figure 2.4). ■

■ **Example 2.4** Your experiment is to roll a standard die twice in a sequence. You record the outcome as a pair; e.g., $(2,5)$ means you got a 2 on the first roll and a 5 on the second. How many different outcomes are possible?

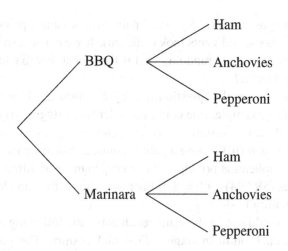

Figure 2.4: Two tasks performed in a sequence that illustrate Theorem 2.3.1.

We can think of the experiment being performed in stages. Stage 1: Roll once. This can produce 6 possible outcomes. Stage 2: Roll again. This can also produce 6 possible outcomes. By the multiplication principle, there are $6 \cdot 6 = 36$ possible outcomes for the experiment. The sample space Ω is given below.

$$\Omega = \{(1,1),(1,2),(1,3),(1,4),(1,5),(1,6),$$
$$(2,1),(2,2),(2,3),(2,4),(2,5),(2,6),$$
$$(3,1),(3,2),(3,3),(3,4),(3,5),(3,6),$$
$$(4,1),(4,2),(4,3),(4,4),(4,5),(4,6),$$
$$(5,1),(5,2),(5,3),(5,4),(5,5),(5,6),$$
$$(6,1),(6,2),(6,3),(6,4),(6,5),(6,6)\}.$$

To view the outcome of our experiment as a sequence of outcomes from each stage and then apply the multiplication principle is very helpful when we need to find the probability of an event A using Definition 2.3.1 (we will call that Method 1).

An equivalent approach is to determine the probability $P(A)$ as the product of the probabilities from each stage (we will call that Method 2).

To illustrate the two methods, we now present a sequence of examples and solve each of them both ways. We also note some differences between the examples that will be formalized later.

■ **Example 2.5** Suppose we have 5 shirts, 3 pairs of shoes and 4 pairs of pants (assume all shirts, shoes, and pants look different). If we chose a shirt, a pair of shoes, and a pair of pants at random, what is the probability to end up with any specific desired outfit?

Method 1. To answer the question using Definition 2.3.1, we need to calculate $|\Omega|$. In this example, one could consider three stages. (i) choose a shirt – this can be done in 5 different ways; (ii) choose a pair of shoes, which can be done 3 ways, and (iii) choose a pair of pants, a choice we can make in 4 ways. By the multiplication principle, the overall number of different outfits we can assemble is $(5)(3)(4) = 60 = |\Omega|$. Each specific outfit is an elementary event, so its probability is $\frac{1}{|\Omega|} = \frac{1}{60}$.

Method 2. We could obtain the same result using the following argument. We can assemble each outfit in stages. First pick a shirt. The probability of picking a specific shirt at random is $\frac{1}{5}$. Then chose a pair of shoes. The probability of selecting the desired pair at random is $\frac{1}{3}$. Finally, chose pants at random. The pair we want will be chosen with probability $\frac{1}{4}$. Now we multiply those probabilities to calculate the probability of the outfit and obtain

$$\frac{1}{5} \cdot \frac{1}{3} \cdot \frac{1}{4} = \frac{1}{60},$$

just as we computed earlier.   ■

■ **Example 2.6** A club has 24 members and needs to appoint a president, vice president, secretary, and treasurer. Alice, Bob, Catherine, and David are members of the club. If appointments to each post are made at random, what is the probability that Alice will be appointed president, Bob – vice president, Catherine – secretary, and David – treasurer?

Method 1. To find the probability using Definition 2.3.1, we need to count how many different appointments are possible.

We break the selection process into stages. First, select president; follow with selecting vice president; next, select a treasurer, and, finally, select a secretary. A president can be selected at random in 24 ways. Once a president is selected, there are 23 choices left for vice president. So, a president and vice president can be selected in $24 \cdot 23$ ways. Similarly, after president and vice presidents are selected, there are 22 possible choices for secretary. Finally, when the first three posts are filled, there will be 21 members left, from which to select a treasurer. By the multiplication principle, the overall number of different ways to fill the leadership posts is

$$|\Omega| = 24 \cdot 23 \cdot 22 \cdot 21.$$

Thus the probability that Alice will be appointed president, Bob – vice president, Catherine – secretary, and David – treasurer is

$$\frac{1}{|\Omega|} = \frac{1}{24 \cdot 23 \cdot 22 \cdot 21} = \frac{20!}{24!}.$$

The representation on the right uses the so-called factorial notation (see Section 6.4 of the Appendix).

Method 2. We can obtain the same result by multiplying the probabilities for each stage. The probability that Alice will be selected for president is $\frac{1}{24}$. Now that Alice has already been selected, there are 23 club members to choose from for vice president, and Bob is among them. The probability that he will be selected is $\frac{1}{23}$. Similarly, the probability that Catherine will be selected to serve as treasurer is now $\frac{1}{22}$, and, finally, the probability that David will fill the post of secretary is $\frac{1}{21}$. Thus, the probability that Alice will be appointed president, Bob – vice president, Catherine – secretary, and David – treasurer is

$$\frac{1}{24} \cdot \frac{1}{23} \cdot \frac{1}{22} \cdot \frac{1}{21} = \frac{1}{24 \cdot 23 \cdot 22 \cdot 21} = \frac{20!}{24!},$$

just as we computed before. ∎

The next theorem is a synopsis of what we have observed in the last two examples.

> **Theorem 2.3.2** If an event A can be considered as occurring in k stages, where the first stage has probability p_1, the second – probability $p_2, \ldots$, and the k-th – probability p_k, then the probability $P(A)$ can be computed as
>
> $$P(A) = p_1 p_2 \cdots p_k.$$

Our next example is different from Example 2.6 in a subtle way.

■ **Example 2.7** A club has 24 members and needs to appoint a committee of four to plan an event. Alice (A), Bob (B), Catherine (C), and David (D) are members of the club. If appointments to the committee are made at random, what is the probability of the event E that Alice, Bob, Catherine, and David will be selected for this committee?

The difference with Example 2.6 is that now the committee does not have designated positions to fill. In Example 2.6 Alice being president and David serving as treasurer is different from Alice being treasurer and David – president. Put differently, if writing ACDB means Alice is president, Catherine is vice-president, David is treasurer, and Bob is secretary, the *order* in which we write the letters matters. This means ACDB is different from DCAB, different from ABCD, and so on. For the example here, such distinction does not exist. We just want to find the probability that the four of them will all be selected to form the committee. Now the committee ABCD is the same as the committee BADC, the same as the committee DBAC, and so on.

We will again think about how to calculate the probability in two different ways.

Method 1. To use Definition 2.3.1, we need to calculate $|\Omega|$. If we start with $24 \cdot 23 \cdot 22 \cdot 21$ as in Example 2.6 we are overcounting because we saw that many different arrangements will result in the same committee. By the multiplication principle, the number of ways ABCD can be arranged is $4 \cdot 3 \cdot 2 \cdot 1 = 4!$ ways. So, these $4!$ options that were different in the context of Example 2.6 produce the *same* committee. Thus, we have overcounted by a factor of $4!$ and, for this example,

$$|\Omega| = \frac{24 \cdot 23 \cdot 22 \cdot 21}{4!} = \frac{24!}{20!4!}.$$

Therefore, the probability of selecting committee with members Alice, Bob, Catherine, and David is

$$P(E) = \frac{1}{|\Omega|} = \frac{1}{\frac{24 \cdot 23 \cdot 22 \cdot 21}{4!}} = \frac{20!4!}{24!}.$$

Method 2. We will now compute $P(E)$ by multiplying the probabilities from each "stage." First, calculate the probability that Alice will be on the team. Picking Alice from the group of 24 has probability $\frac{1}{24}$, and she can hold any of the four appointments. Thus, the probability for Alice to be on the team is $\frac{4}{24}$. Next, let's compute the probability that Bob is on the team. Now that Alice is selected, there are 23 club members to pick from, and the selected person could hold one of the remaining three appointments. So, the probability that Bob is on the team is $\frac{3}{23}$. We continue in the same fashion for the remaining two steps. Thus, the probability we want is

$$P(E) = \frac{4}{24} \cdot \frac{3}{23} \cdot \frac{2}{22} \cdot \frac{1}{21} = \frac{4!}{24 \cdot 23 \cdot 22 \cdot 21} = \frac{20!4!}{24!}.$$

∎

In the last three examples, you may have felt that one of the two methods we illustrated appeared to be more natural to you to use than the other. Even though they are equivalent, a problem may be more easily solved using one approach versus the other, and often this depends on personal preference. We will use these approaches interchangeably from now on, and will come back to this later when we discuss independence and conditional probability.

Examples 2.6 and 2.7 also show that counting arguments we derived from the multiplication principle (Theorem 2.3.1) can be subtle and which one to use in a problem may not be immediately obvious. In the first case, the ranking (order of an element) within the leadership team matters, while in the second case it doesn't. We next generalize the observations we made when counting the number of elements in $|\Omega|$ in each case. However, we caution against trying to memorize these as mathematical formulas that you would immediately go to find a solution. Instead, we recommend that you approach each problem individually and work out a solution, similar to what we did in the examples above.

Theorem 2.3.3

1. The number of ways to select k objects from n objects where order of selection of the objects *does* matter is

$$\frac{n!}{(n-k)!}. \tag{2.1}$$

We say that this gives the number of *permutations* that can be formed of k objects selected from a group of n.

2. The number of ways to select k objects from n objects where order of selection of the objects *does not* matter is

$$\frac{n!}{k!(n-k)!}. \tag{2.2}$$

We say that this gives the number of *combinations* that can be formed of k objects selected from a group of n. The following notation is common for the number of combinations:

$$\binom{n}{k} = \frac{n!}{k!(n-k)!}.$$

To reiterate, we consider permutations when the order matters and combinations when it doesn't.

We conclude this section with two more examples, which we solve by combining some of the techniques presented earlier.

■ **Example 2.8** A group of 50 students has 10 seniors, 12 juniors, and 28 sophomores. We select a group of 8 students at random. What is the probability that we select 3 seniors, 1 junior, and 4 sophomores?

We will use Definition 2.3.1 (Method 1). First, we need to find $|\Omega|$; that is, determine how many ways a group of 8 from 50 can be selected. Second, we need to know how many ways those 8 can be selected to give the desired make up. That is, if A is the event "of the eight selected, 3 are seniors, 1 is a junior, and 4 are sophomores," we will need to find $|A|$.

For $|\Omega|$: We are selecting 8 people from 50. Using arguments similar to those in Example 2.7 and Equation (2.2), this can be done in

$$|\Omega| = \frac{50!}{8!42!} \text{ ways.}$$

For $|A|$: We think of performing the experiment in stages.
Selecting 3 seniors from 10 seniors can be done in

$$\frac{10!}{3!7!} \text{ ways.}$$

Selecting 1 junior from 12 juniors can be done in

$$\frac{12!}{1!11!} \text{ ways.}$$

Finally, selecting 4 sophomores from 28 sophomores can be done in

$$\frac{28!}{4!24!} \text{ ways.}$$

Thus, by the multiplication principle, the number of ways that we can make our selection successfully is

$$|A| = \left(\frac{10!}{3!7!}\right) \cdot \left(\frac{12!}{1!11!}\right) \cdot \left(\frac{28!}{4!24!}\right).$$

Now, using Definition 2.3.1, we obtain

$$P(A) = \frac{|A|}{|\Omega|} = \frac{\left(\frac{10!}{3!7!}\right) \cdot \left(\frac{12!}{1!11!}\right) \cdot \left(\frac{28!}{4!24!}\right)}{\frac{50!}{8!42!}}.$$

■

From Theorem 2.2.1, we know that if A is an event and A^c is its complement, $P(A) = 1 - P(A^c)$. This is useful because, in some cases, computing $P(A^c)$ may be easier than computing $P(A)$ directly. Our next example demonstrates this approach.

■ **Example 2.9** **The Birthday Problem.** There are 10 people in a room. What is the probability that at least two of them have the same birthday?

It is tedious and lengthy to give the event that two people have the same birthday (and very easy to make a mistake). It is much simpler to find the probability that all have *different* birthdays. Notice that these two events are complements of one another.

We will apply the approach where we multiply the probabilities of each "stage" (Method 2). Select people one at a time. Ignoring leap years, the first person has 365 out of 365 days to have a different birthday than all selected previously (namely none). The probability of this happening is $\frac{365}{365}$. The second person will have 364 out of 365 days to have a different birthday than the first person selected. The probability of both occurring is

$$\frac{365}{365} \cdot \frac{364}{365}.$$

Continuing this until 10 people have been selected we have a probability of all having different birthdays is

$$\frac{365}{365} \cdot \frac{364}{365} \cdot \frac{363}{365} \cdots \frac{356}{365} = \frac{(365 \cdot 364 \cdot 363 \cdots 356)}{365^{10}}.$$

Now, the probability that at least two people will have the same birthday is

$$1 - \frac{(365 \cdot 364 \cdot 363 \cdots 356)}{365^{10}} \approx 0.117.$$

■

Exercise 2.1 Consider a set Ω with n elements; that is, $|\Omega| = n$. How many different subsets does A have? That is, how many different events are possible for an experiment with sample space Ω? *Hint.* Use the multiplication principle. Looking at each element of Ω, you have two choices, to put that element in the subset or not. So, two choices for the first element (Stage 1), two choices for the second element (Stage 2), and so on. ■

Exercise 2.2 A small company has 12 employees. Its president wants to assign specific tasks to three of them. Each task requires that the worker puts in some overtime, for which they will be compensated at $30 per hour. Task 1 requires 3 hours of overtime, Task 2 – 1 hour, and Task 3 – 5 hours of overtime. The selection is made at random. If Alice, Ben, and Cathy work at the company, what is the probability that they will be chosen for the tasks and Alice will receive $90, Ben will receive $30, and Cathy will receive $150 in compensation for their overtime effort?

Exercise 2.3 A small company has 12 employees. Its president wants to assign specific tasks to three of them. Each task requires that the worker puts in 3 hours of overtime, for which the workers will be compensated at $30 per hour. The selection is made at random. If Alice, Ben, and Cathy work at the company, what is the probability that they will be chosen to perform these tasks?

Exercise 2.4 If there are 15 people in a room, what is the probability that 2 of them have the same birthday?

Exercise 2.5 We have a club of 15 people – 9 women and 6 men. We select 5 for a committee. What is the probability the committee consists of 2 women and 3 men? What is the probability the committee consists of 5 men?

Exercise 2.6 We are going to flip coins in groups of 5.
 1. For a particular group, what is the probability that all 5 in the group are heads?
 2. If we do the flipping with 20 such groups, what is the probability that none of the groups have all 5 heads?
 3. What is the probability that at least one of the twenty groups has all heads?

Exercise 2.7 We have an urn that has 7 green, 5 red, and 12 blue balls. We draw 2 balls without replacement. What is the probability we get exactly 1 green and 1 blue ball? *Hint.* Possible cases of the 2 balls: 1 green and 1 blue, 1 green and 1 red, 1 red and 1 blue.

Exercise 2.8 Jack and Jill roll a die. Jill goes first. Jill wins if she gets a 1 or 2; Jack wins if he gets a 3, 4, or 5. No one wins if the roll is 6. What is the probability that Jack wins on his first or second roll?

Exercise 2.9 (Like the birthday problem.) If we roll a 12-faced dice 8 times, what is the probability some number appears more than once? ∎

Exercise 2.10 There are n tickets in a lottery, of which m are winners. How large is the probability of a win for a person holding k tickets? ∎

Exercise 2.11 In how many different ways can 10 students be seated in a classroom with 30 desks? *Hint.* Think of this in stages. First, find how many ways you can select 10 desks from the 30. Second, find how many different ways the students can be arranged among the 10 desks. Then, use the multiplication principle. ∎

Exercise 2.12 Cards are dealt one after the other from a standard deck of cards without replacement. Calculate the probability that
1. the first two cards dealt will both be diamonds;
2. the third card will be the first diamond;
3. the first card will be red and the second card will be black;
4. of the first two cards dealt, one will be a diamond and the other black;
5. the first four cards dealt will be red.
  ∎

Exercise 2.13 A committee of 4 is chosen at random from 5 couples. What is the probability that the committee will not include a person and their partner? ∎

Exercise 2.14 An urn contains 25 balls – 8 red, 15 green, and 2 red. Three balls are drawn at random in a sequence without replacement. Find the probability that
1. a red, a green, and a blue ball will be drawn, in that order;
2. the first two balls are green;
3. the first ball is blue and the third ball is green.
  ∎

> Exercise 2.15 Consider again the birthday problem from Example 2.9. With 23 people in the room, find the probability that at least two of them have the same birthday. Do you find the result surprising? ∎

2.4 Conditional Probability

In some of the examples from the previous section, we already used conditional probabilities implicitly. Here, we formalize the concept.

In conditional probability, we consider two events, say A and B, from a probability space. Knowing that the event B has occurred, we want to find the probability that A occurs. The symbol for this is $P(A|B)$. This is read "the probability of A, given B."

∎ **Example 2.10** We draw a card from a standard deck. We draw one card and want to know the probability that it is a club (event A). Absent any other information, we will have

$$P(A) = \frac{13}{52} = \frac{1}{4},$$

since there are 13 clubs in the standard deck.

Suppose, however, that before we see the card we have drawn, someone gives us the additional information that it's a black card (event B). Now, knowing the event B has occurred *eliminates part of the sample space*. In this example, the sample space is down to the 26 black cards. We are now asking, from those 26 cards, what is the probability that the card drawn was a club. Since there are 13 clubs in the group of 26, the probability in this case is $\frac{13}{26} = \frac{1}{2}$.

Symbolically, we express this by writing

$$P(A|B) = \frac{1}{2}$$

∎

Thus, having the additional information that the event B has occurred, changed the probability of the event A.

∎ **Example 2.11** In Weathertown, the weather each day is either sunny or rainy with equal probability, except on Sundays, when the weather is always sunny. You know you will be vacationing there for three consecutive days, and your three days will cover Sunday; that is, you will have at least one sunny day (event B). You are yet uncertain about which other two days you will be there. What is the probability that it will be sunny for the three days you are there (event A)?

The sample space here is $\Omega = \{SSS, SSR, SRS, SRR, RSS, RSR, RRS, RRR\}$, where we use S to mean a sunny day and R to mean a rainy day.

Knowing that one of the three days of your vacation (Sunday) will be sunny, the outcome RRR is no longer possible and can be eliminated from the sample space. This reduces the sample space to $\{SSS, SSR, SRS, SRR, RSS, RSR, RRS\}$. Thus, the probability that you will have sunny weather for your three day vacation in Weathertown is $\frac{1}{7}$.

Symbolically, we write $P(A|B) = \frac{1}{7}$.   ■

Notice that for our last example, we can write

$$P(A|B) = \frac{1}{7} = \frac{1}{8} \cdot \frac{8}{7},$$

and give interpretation of the two fractions as probabilities in the following way: $\frac{1}{8}$ is the probability $P(A \cap B)$ – the probability that all three days are sunny, and $\frac{7}{8}$ is the probability $P(B)$ that there will be at least one sunny day (Sunday). Thus, we can write

$$P(A|B) = \frac{P(A \cap B)}{P(B)}.$$

This may seem as an unnecessary over-complication, considering how straightforward our argument was. However, it provides context for the general definition.

Definition 2.4.1 Let A and B be two events. Then the *conditional probability*, of A given B, denoted by $P(A|B)$, is

$$P(A|B) = \frac{P(A \cap B)}{P(B)}, \text{ provided } P(B) \neq 0. \qquad (2.3)$$

Equation (2.3) can be rewritten as

$$P(A \cap B) = P(B)P(A|B). \qquad (2.4)$$

It states that the probability that A and B occur is the product of the probability that B occurred, multiplied by the probability of A, *knowing that B occurred*.

You may have noticed that we used this approach several times already without referring to any formulas, just using the reduced sample space (see, e.g., Examples 2.6 and 2.9). We give one more such example here.

■ **Example 2.12** Two cards are drawn from a standard deck of cards without replacement. Find the probability that the first card is the ace of spades and the second card is a black card.

The "experiment" here has two clear stages: First, draw the ace of spades (event B). Second, without putting the ace of spades back into the deck, draw a black card (event A). The probability to draw the ace of spades first is $P(B) = \frac{1}{52}$. Once we know that, the conditional probability to draw a black card is $P(A|B) = \frac{25}{51}$, since there are now 51 cards left in the deck, of which 25 are black. So, we have

$$P(A \cap B) = P(B)P(A|B) = \frac{1}{52} \cdot \frac{25}{51}.$$

∎

Notice that if we interchange the roles of A and B in Definition 2.4.1, we get

$$P(B|A) = \frac{P(B \cap A)}{P(A)} = \frac{P(A \cap B)}{P(A)},$$

which leads to

$$P(A \cap B) = P(A)P(B|A). \tag{2.5}$$

Combining Equations (2.4) and (2.5), gives

$$P(A \cap B) = P(B)P(A|B) = P(A)P(B|A). \tag{2.6}$$

In other words, it tells us that when we break the outcome of an experiment into stages, the order in which we perform the stages to compute its probability is not important. We can choose the order that makes computing the probability of the stages easier.

The product-of-probabilities rule from Equation (2.6) extends to more than two sets. The theorem below is the more rigorous version of Theorem 2.3.2, which we presented and used earlier without giving a proof.

Theorem 2.4.1 If A, B, C are all events with nonzero probability, then

$$P(A \cap B \cap C) = P(A)P(B|A)P(C|A \cap B).$$

Here $P(C|A \cap B)$ is the probability of C, knowing that both A and B have occurred.

For four events A, B, C, and D,

$$P(A \cap B \cap C \cap D) = P(A)P(B|A)P(C|A \cap B)P(D|A \cap B \cap C), \text{ and so on.}$$

Proof. We will present the proof for three events. Establishing the result for four events is left as an exercise (see Exercise 2.17).

Using Equation (2.4), we have

$$P(A \cap B \cap C) = P((A \cap B) \cap C) = P(A \cap B)P(C|A \cap B)$$
$$= P(A)P(B|A)P(C|A \cap B).$$

∎

Exercise 2.16 Reconsider Example 2.11, but now assume you have decided that Sunday will be the last day of your three-day vacation. What is the probability that you will be able to enjoy sunshine during your entire stay? Do the problem two different ways: (1) by using the reduced sample space and (2) by multiplying probabilities. ▪

Exercise 2.17 Prove that if A, B, C, and D are events with nonzero probabilities, then

$$P(A \cap B \cap C \cap D) = P(A)P(B|A)P(C|A \cap B)P(D|A \cap B \cap C).$$

Revisit Example 2.6 and explain what $P(A)$, $P(B|A)$, $P(C|A \cap B)$ and $P(D|A \cap B \cap C)$ are in the context of that example. ▪

Exercise 2.18 Suppose a box has 3 red marbles and 2 black ones. We select 2 marbles. What is the probability that second marble is red given that the first one is red? ▪

Exercise 2.19 There is a crowd of 30 people, 20 men and 10 women, one of whom is your sister. A friend selects a person at random and it is a woman. What is the probability it is your sister? ▪

Exercise 2.20 The probability of a student passing in science is 3/5 and the probability of a student passing in both science and math is 1/4. What is the probability that a student passes in math, knowing that they passed in science? ▪

Exercise 2.21 You roll a die. An odd number comes up. What is the probability it is a prime number? (Remember, 1 is not a prime!) ▪

2.5 Independence

Sometimes, as our next example shows, $P(A|B)$ may be the same as $P(A)$.

▪ **Example 2.13** We draw one card from a standard deck of cards. Suppose now we know that the card drawn was red (event B), and we want to know the probability that the card was an ace (event A).

Since there are four aces in the deck, the probability to draw an ace is

$$P(A) = \frac{4}{52}.$$

Since there are 2 red aces and 26 red cards, the conditional probability

$$P(A|B) = \frac{2}{26} = \frac{4}{52} = P(A).$$

So, we get that $P(A|B)$ is the same as the probability that a card drawn from the deck is an ace, knowing nothing about the color of the card in advance. ▪

When $P(A|B) = P(A)$, the additional information we get from knowing that event B occurred is not affecting the probability of A. We can then think of the event A as being independent from the event B. In such cases, Equation (2.4) becomes

$$P(A \cap B) = P(A)P(B).$$

These considerations motivate the following definition.

Definition 2.5.1 Two events A and B are called *independent events* when

$$P(A \cap B) = P(A)P(B). \tag{2.7}$$

Definition 2.5.1 tells us that if we know that two events are independent, the probability that both of them occurred is obtained by multiplying the probabilities of each event.

Exercise 2.22 In the manufacturing of a certain motorcycle part, defects of one type occur with probability 0.1 and defects of a second type with probability 0.05. A part is selected at random from the production line. Assuming that the two types of defects are independent, find the probability that

1. the part does not have both kinds of defects;
2. the part is defective (that is, it has at least one defect);
3. the part has only one defect.

Exercise 2.23 A card is drawn from a standard deck of cards. Is the event to draw a club independent from the event to draw an ace? Use Definition 2.5.1 to justify your answer.

Exercise 2.24 You roll a fair die twice. Suppose A is the event that you roll a 3 on the first roll and an odd number on the second roll. Suppose B is the event that you rolled a sum of at least 10 points from the two rolls. Are the events A and B independent? Use Definition 2.5.1 to justify your answer.

Exercise 2.25 A fair coin is flipped three times. Let A be the event that you obtained at least 1 heads, and B be the event that the second and the third flip were tails. Are the events A and B independent? Use Definition 2.5.1 to justify your answer.

Exercise 2.26 If A and B are independent events, prove that A^c and B^c are also independent. *Hint.* Use Definition 2.5.1 and that $P(A^c) = 1 - P(A)$ (Theorem 2.2.1).

2.6 The Bayes' Theorem

We discuss our next topic by way of an example. The theorem that relates the method is called the Bayes' Theorem, and is tedious to memorize. However, in most problems, the solution is intuitive, and trying to apply the theorem from memory is not a good idea. We begin with an example to illustrate that before we state the theorem.

■ **Example 2.14** Consider three containers, about which we know the following:

- Container A contains 4 red and 6 green balls;
- Container B contains 4 red and 2 green balls, and
- Container C contains 10 red and 8 green balls.

A container is chosen at random, and a ball is drawn from that container. Container A has a probability of 1/2 of being chosen, container B has a probability of 1/3 of being chosen and container C has a probability of 1/6 of being chosen. We do not know which container was chosen, but we are told that the selected ball is green. What is the probability it came from container B?

The tree diagram in Figure 2.5 gives a convenient way to solve the problem.

Let $P(A)$, $P(B)$, and $P(C)$ denote the respective probabilities that containers A, B, and C are selected. Let $P(G)$ denote the probability that the ball drawn is green.

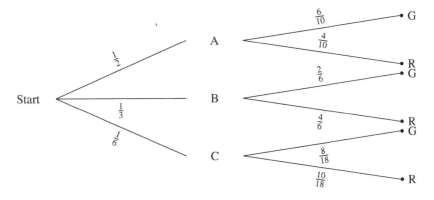

Figure 2.5: Tree diagram for Example 2.14.

We know, from the definition of conditional probability, that

$$P(B|G) = \frac{P(B \cap G)}{P(G)}.$$

From the tree diagram in Figure 2.5, we see that $P(B \cap G) = \frac{1}{3} \cdot \frac{2}{6}$. The probability $P(G)$ can be calculated by looking at the different possible paths on the tree to get from "Start" to a G.

$$P(G) = \left(\frac{1}{2}\right)\left(\frac{6}{10}\right) + \left(\frac{1}{3}\right)\left(\frac{2}{6}\right) + \left(\frac{1}{6}\right)\left(\frac{8}{18}\right).$$

Thus,

$$P(B|G) = \frac{P(B \cap G)}{P(G)} = \frac{\left(\frac{1}{3}\right)\left(\frac{2}{6}\right)}{\left(\frac{1}{2}\right)\left(\frac{6}{10}\right) + \left(\frac{1}{3}\right)\left(\frac{2}{6}\right) + \left(\frac{1}{6}\right)\left(\frac{8}{18}\right)} \approx 0.23.$$

■

Another way to look at the same problem is depicted in Figure 2.6. We can draw the sample space Ω as the union $\Omega = A \cup B \cup C$, where the sets A, B, and C are disjoint. In Example 2.14, they represent all possible outcomes coming from container A, B, and C. The event $G =$ "A green ball is drawn" is depicted by the ellipse inside Ω.

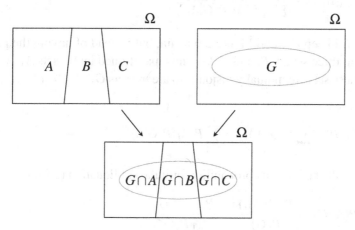

Figure 2.6: A more general illustration of the Bayes' Theorem for Example 2.14.

If we know (as we did in Example 2.14) the probabilities $P(A), P(B)$, and $P(C)$ and the probability of G when each of the events A, B, and C occurred (that is, if we also know $P(G|A), P(G|B)$, and $P(G|C)$), we get to write

$$P(G) = P(G \cap A) + P(G \cap B) + P(G \cap C)$$
$$= P(A)P(G|A) + P(B)P(G|B) + P(C)P(G|C).$$

Now, if we want the probability $P(B|G)$, we have

$$P(B|G) = \frac{P(B \cap G)}{P(G)}$$
$$= \frac{P(B \cap G)}{P(G \cap A) + P(G \cap B) + P(G \cap C)}$$
$$= \frac{P(B \cap G)}{P(A)P(G|A) + P(B)P(G|B) + P(C)P(G|C)}.$$

The result we obtained is a special case of the following theorem.

Theorem 2.6.1 (Bayes) Let events $A_1, A_2, \ldots, A_n$ be disjoint and $\Omega = A_1 \cup A_2 \cup \cdots \cup A_n$. Let G be any event. Then

$$P(A_i|G) = \frac{P(A_i)P(G|A_i)}{\sum_{i=1}^{n} P(A_i)P(G|A_i)}. \tag{2.8}$$

Proof. We keep in mind Figure 2.6 again, but instead of having the partition of Ω into three sets A, B, and C, Ω is now partitioned into n sets $A_1, A_2 \ldots, A_n$. Since these sets are mutually disjoint, so are the sets $G \cap A_1, G \cap A_2, \ldots, G \cap A_n$. Thus

$$P(G) = \sum_{i=1}^{n} P(G \cap A_i) = \sum_{i=1}^{n} P(A_i)P(G|A_i). \tag{2.9}$$

From the conditional probability formula and Equation (2.9), we now have

$$P(A_i|G) = \frac{P(G \cap A_i)}{P(G)} = \frac{P(A_i)P(G|A_i)}{\sum_{i=1}^{n} P(G \cap A_i)},$$

which completes the proof.   ∎

Equation (2.8) in Bayes' Theorem may seem quite intimidating at first. It is a good example for how the mathematical notation may sometimes obscure the straightforward logic behind the result. We recommend that you do NOT try to memorize this equation. Instead, rely on basic principles, just as we did in Example 2.14.

■ **Example 2.15** You live in a town of 100,000 inhabitants, some of whom are infected by a virus.There is a test to determine if an individual is infected. If an individual is infected, the test is correct every time, but if an individual is not infected, the test gives a false positive 5% of the time. Suppose you know 40% of the population is infected. You take the test and test positive. How worried should you be?

We will solve this example in two different ways. The "easy" solution uses our basic understanding of computing probabilities. The "complicated solution" will be to apply the formula from Equation (2.8). We do this to demonstrate that each solution will lead to the same answer and underscore the differences in complexity.

The easy solution: You know there are 40,000 sick people, and each one of them will test positive. You also know that 5% of the 60,000 who are not infected (that is 3000 people) will also get a positive reading. Thus there are $40,000 + 3000 = 43,000$ positive tests. The question is: Are you in the group of $40,000$ or the group of 3000. The probability that you have the virus (i.e., the probability you are in the group of $40,000$) is $40,000/43,000$ or approximately 93%. Thus, unfortunately, you should be pretty worried.

The "complicated" *solution*: We break the sample space Ω into two sets: The set S of sick people and the set H of healthy people. We have $S \cap H = \varnothing$ and $\Omega = S \cup H$, so Theorem 2.6.1 applies. Let $+$ denotes the set of those with positive tests.

We know the following:

- $P(+|S) = 1$; that is, for sick people, the test is correct every time.
- $P(+|H) = 0.05$; that is, we get false-positives with probability 0.05.
- $P(S) = \frac{40,000}{100,000} = 0.4$.
- $P(H) = \frac{60,000}{100,000} = 0.6$.

We need to compute $P(S|+)$. From Equation (2.8), we obtain

$$P(S|+) = \frac{P(S)P(+|S)}{P(S)P(+|S) + P(H)P(+|H)}$$
$$= \frac{(0.4)(1)}{(0.4)(1) + (0.6)(0.05)}$$
$$= \frac{0.4}{0.43} \approx 0.93.$$

We obtained the same answer, but, note again, that the "easy" intuitive solution is preferable.

∎

■ **Example 2.16** In class Rocket Science, 80% of the students do the homework regularly and 20% don't. The probability of passing the class when doing homework regularly is 90%, and the same probability is 50% when that's not the case. A student from the class is selected at random.

1. What's the probability that the student passes the class?
2. The student selected at random passes the exam. What's the probability that they didn't do the homework regularly?

Let X denote the event that a student does homework regularly, and Y be the event that they don't. Let Z denote the event that a student passes the class. We know $P(X) = 0.8$, $P(Y) = 1 - 0.8 = 0.2$, $P(Z|X) = 0.9$, and $P(Z|Y) = 0.5$.

(1) We want to find $P(Z)$. Since $X \cap Y = \varnothing$, and $X \cup Y = \Omega$, we have that the events $Z \cap X$ and $Y \cap X$ are also disjoint and

$$
\begin{aligned}
P(Z) &= P(Z \cap X) + P(Z \cap Y) \\
&= P(X)P(Z|X) + P(Y)P(Z|Y) \\
&= (0.8)(0.9) + (0.2)(0.5) \\
&= 0.82
\end{aligned}
$$

(2) Here, we need to find $P(Y|Z)$. From the conditional probability formula, and using the result from part (1),

$$
P(Y|Z) = \frac{P(Y \cap Z)}{P(Z)} = \frac{P(Z \cap Y)}{P(Z)} = \frac{P(Y)P(Z|Y)}{P(Z)} = \frac{(0.2)(0.5)}{0.82} \approx 0.122.
$$

∎

Exercise 2.27 Suppose that a test for a rare disease sometimes makes mistakes: 1 in 100 of those free of the disease have positive test results, and 2 in 100 of those having the disease have negative test results. The rest are correctly identified. One person in 1000 has the disease. Find the probability that a person with a positive test has the disease.

Exercise 2.28 We have two urns. Urn 1 has 2 red and 5 green balls. Urn 2 has 3 red and 10 green balls. An urn is selected and a ball is drawn. It is a red ball. What is the probability urn 1 was selected?

Exercise 2.29 Maria has 2 coins in her pocket. One is a fair coin and the other has two heads.
1. Maria picks a coin at random and flips it. It is heads. What is the probability it is the fair coin?
2. Maria flips the coin twice and gets heads both times. What is the probability it is the fair coin?

Exercise 2.30 A bag contains three coins, one of which is fake and has two heads, while the other two coins are ordinary and not biased. A person chooses a coin from the bag at random and tosses it four times in a row. If that person tells you that heads turned up each time, what is the probability that this is the two-headed coin?　　　　■

Exercise 2.31 In a town, there are 40% Democrats and 60% Republicans, 30% women and 70% men. You choose a voter at random who voted Democrat. What is the probability you chose a woman?　　　■

7　Random Variables and Expected Value

In cases when we are interested in a certain numerical value coming out of a random experiment, it is helpful to use random variables.

Definition 2.7.1 A random variable is a function $X : \Omega \to \mathbb{R}$. For each elementary event from the sample space Ω, X assigns a numerical value.

To understand what this definition tells us, recall that the individual elements of a sample space Ω are called elementary events or data points of the sample space. When we toss a single coin two times, the sample space is $\Omega = \{HH, HT, TH, TT\}$, so the elementary events are $\{HH\}$, $\{HT\}$, $\{TH\}$, $\{TT\}$. When we draw two cards from a standard deck without replacement, we know that there are $52 \cdot 51$ possible outcomes. Thus Ω has $52 \cdot 51$ elements. Each one of those is an elementary event that looks like $\{3\heartsuit, K\spadesuit\}$, $\{A\clubsuit, 9\diamondsuit\}$, $\{5\diamondsuit, 7\heartsuit\}$, and so on.

Now, let's consider examples of random variables defined for these experiments.

■ **Example 2.17** You flip a coin twice (this is the experiment). Consider the random variable X that assigns to each possible outcome (elementary event) the number of tails you get in an experiment. We have no tails for the outcome $\{HH\}$, one tail for the outcomes $\{HT\}$, $\{TH\}$, and two tails for the outcome $\{TT\}$. So, according to the definition above, the function X is defined as $X(\{HH\}) = 0$, $X(\{HT\}) = 1$, $X(\{TH\}) = 1$, and $X(\{TT\}) = 2$.　　■

■ **Example 2.18** You draw two cards from a standard deck without replacement. Consider the random variable X that assigns to each of the $52 \cdot 51$ possible outcomes the sum of points you will get, counting an ace as 1 point, a jack as 11 points, a queen as 12 points, and a kings as 13 points. Then we will have, e.g., $X(\{A\clubsuit, 9\diamondsuit\}) = 10$, $X(\{3\heartsuit, K\spadesuit\}) = 16$, $X(\{5\diamondsuit, 7\heartsuit\}) = 12$, and so on.　　■

A random variable may also have infinitely many values, as our next example shows.

■ **Example 2.19** In the game of Trouble, one must roll a 6 on a dice before they can move around the board. Let X be the random variable the value of which is the number of the roll on which you got a 6 for the first time.

If your first roll is a 6, $X = 1$. If the first six for you is your second roll, then $X = 2$. If the first six for you happens when you roll for a thousandth time, $X = 1000$, and so on. In theory, you may keep rolling many, many times, and still not get a 6. Thus the set of possible values for this random variable is the set of all positive integers $\mathbb{N} = \{1, 2, 3, \dots\}$.

We will return to this example in Section 2.8.2 when we discuss the so-called Geometric Distribution.   ■

It is helpful to know the probability with which a random variable takes a certain value. To find them, we do the same thing as in our last two examples but in a slightly different order. First, we decide what values the random variable of interest can take. Second, we decide which elementary events produce that outcome.

■ **Example 2.20** In example 2.17, recall that the possible values for the random variable X that gives the number of tails from two coin flips are 0, 1, and 2. For each value, we now decide which elementary events produce that value. This is summarized in Table 2.1. Notice that the probabilities in the last column sum up to 1, as X can only take one of the values listed.   ■

Value x of X	Elementary events with that value	$P(X = x)$
0	$\{HH\}$	$\frac{1}{4}$
1	$\{HT\}, \{TH\}$	$\frac{1}{4} + \frac{1}{4} = \frac{1}{2}$
2	$\{TT\}$	$\frac{1}{4}$

Table 2.1: Table for the random variable from Example 2.20.

Note that in Example 2.20, all outcomes are equally likely. Thus $P(X = 1) = \frac{1}{4} + \frac{1}{4} = \frac{1}{2}$. If A is the event "exactly one tail from the two coin flips occurred," we can also use Definition 2.3.1 to compute this probability as

$$P(X = 1) = P(A) = \frac{|A|}{|\Omega|} = \frac{2}{4} = \frac{1}{2}.$$

Therefore, when we look at random variables for experiments with equally likely outcomes, we can compute $P(X = x)$ for a certain outcome x by counting the number of elementary events that can produce this outcome. Also, since in most cases of interest, the size of the sample space is large, we usually do not list all elementary outcomes that produce the desired value of the random variable. That is, we would usually give a table like Table 2.1 but without the middle column. Our next example does that.

■ **Example 2.21** Two cards are drawn (without replacement) from a standard deck of cards. Let X be the random variable that gives the number of red cards from the two drawn. Then X can take values $0, 1$, or 2.

The value $X = 0$ occurs when both cards are black. This happens with probability

$$P(X = 0) = \frac{26 \cdot 25}{52 \cdot 51} = \frac{1}{2} \cdot \frac{25}{51},$$

because we have probability $\frac{26}{52}$ for the first card to be black and probability $\frac{25}{51}$ for the second card to be black (after a black on the first draw).

The value $X = 1$ occurs when one red and one black card are drawn. The probability that the first card is red and the second black is $\frac{26 \cdot 26}{52 \cdot 51}$ (draw a black card first with probability $\frac{26}{52}$, then draw a red with probability $\frac{26}{51}$). Similarly, the probability that the first card is red and the second is black is also $\frac{26 \cdot 26}{52 \cdot 51}$. Thus,

$$P(X = 1) = \frac{26 \cdot 26}{52 \cdot 51} + \frac{26 \cdot 26}{52 \cdot 51} = 2 \cdot \frac{26 \cdot 26}{52 \cdot 51} = \frac{26}{51}$$

Finally, following the same logic,

$$P(X = 2) = \frac{26 \cdot 25}{52 \cdot 51} = \frac{1}{2} \cdot \frac{25}{51}.$$

This information in summarized in a simplified format in Table 2.2. Notice that the probabilities in the second column sum up to 1.   ■

x	$P(X = x)$
0	$\frac{1}{2} \cdot \frac{25}{51}$
1	$\frac{26}{51}$
2	$\frac{1}{2} \cdot \frac{25}{51}$

Table 2.2: Distribution table for the random variable X in Example 2.21.

Each of Tables 2.1 and 2.2 gives the *probability distribution* for the respective random variable X. More formally, we have the following definition.

Definition 2.7.2 Let X be a random variable with finitely many possible outcomes $x_1, x_2, \ldots, x_n$. If $p_1, p_2, \ldots, p_n$ are numbers such that

$$P(X = k) = p_k = p(k), \quad k = 1, 2, \ldots, n,$$

and $p_1 + p_2 + \cdots + p_n = 1$, we say that the probabilities $p_1, p_2, \ldots, p_n$ give the *probability distribution* of X.

As we will see in Section 2.8.2, this definition generalizes to the case when a random variable can take infinitely many values, when it is clear how to interpret the infinite sum $p_1 + p_2 + \cdots + p_n + \cdots = 1$.

It is often helpful to know what the "average" value of a random variable is. Computing such averages extends the well-known process of taking the average for a set of numbers.

■ **Example 2.22** Suppose you take 8 tests in your Probability class and get the following scores:

$$80, 70, 80, 90, 70, 70, 90, 70.$$

Then your average test score is

$$\frac{80 + 70 + 80 + 90 + 70 + 70 + 90 + 70}{8}$$
$$= \frac{70 + 70 + 70 + 70 + 80 + 80 + 90 + 90}{8}$$
$$= \frac{4(70) + 2(80) + 2(90)}{8}$$
$$= 70\left(\frac{4}{8}\right) + 80\left(\frac{2}{8}\right) + 90\left(\frac{2}{8}\right)$$
$$= 77.5.$$

Let's focus on the expression

$$70\left(\frac{4}{8}\right) + 80\left(\frac{2}{8}\right) + 90\left(\frac{2}{8}\right).$$

What we have done here is to multiply each of your scores by the fraction of tests for which you got that score and summed the values.   ■

■ **Example 2.23** Suppose a group of students took a test for a maximum score of 100. You know that 5% of the students scored 50, 20% scored 65, 45% scored 75, 15% scored 85, 10% scored 95, and 5% scored 100. Then, following the principle above, the average of the group will be

$$(0.05)50 + (0.2)65 + (0.45)75 + (0.15)85 + (0.1)95 + (0.05)100 = 76.5$$

∎

Notice that in the two examples above, no one made the average score. Also, in the second problem, the actual number of people who took the test is not relevant.

The next definition generalizes these two examples.

Definition 2.7.3 If X is a discrete random variable and $p(x) = P(X = x)$, then the *expected value* of X, denoted $\mathbb{E}(X)$, is calculated to be

$$\mathbb{E}(X) = \sum_x xp(x).$$

The sum is over all possible values of the random variable X.

∎ **Example 2.24** Imagine you are playing a game where you roll a fair die once. If you

roll a 1, you win $10,
roll a 2, you win $4,
roll a 3, you lose $2,
roll a 4, you lose $3,
roll a 5, you lose $4,
roll a 6, you lose $1.

Let X be the random variable whose values give how much you have won on a roll. Thus the possible values for X are $10, 4, -2, -3, -4, -1$, where negative values represent losses. If you play that game many times, what would you expect to win/lose on average? That would be the expected value of the random variable X. Since all outcomes from the roll have equal probability $\frac{1}{6}$, the expected value of X in this case is

$$\mathbb{E}(X) = \frac{1}{6}(10) + \frac{1}{6}(4) + \frac{1}{6}(-2) + \frac{1}{6}(-3) + \frac{1}{6}(-4) + \frac{1}{6}(-1) = \$0.67.$$

So, in a long run, if you play that game many, many times, your average gain would be 0.67 (provided, of course, that you are willing to take the risk to gamble in the first place). ∎

∎ **Example 2.25** Calculate the expected value of the random variable with distribution given in Table 2.2. We calculated those probabilities in Example 2.21.

In this case we have

$$\mathbb{E}(X) = 0 \cdot P(X = 0) + 1 \cdot P(X = 1) + 2 \cdot P(X = 2)$$
$$= 0 \cdot \left(\frac{1}{2} \cdot \frac{25}{51} \right) + 1 \cdot \left(\frac{26}{51} \right) + 2 \cdot \left(\frac{1}{2} \cdot \frac{25}{51} \right)$$
$$= \frac{25}{51} + \frac{26}{51}$$
$$= 1.$$

Recall that the random variable X represented how many red cards you will draw from a standard deck without replacement. The expected value we just calculated, $\mathbb{E}(X) = 1$, should make intuitive sense: If we repeat the experiment of drawing two cards many times, we would expect, on average, to draw one red and one black card. ∎

Exercise 2.32 You flip a fair coin three times. What is the expected value of the number of heads observed?

Exercise 2.33 You pay \$2 to play the following game: Draw a ball from an urn and look at the number on the ball. You win the amount in dollars equal to that number. Find the expected value of your net gain if:
1. The urn has 4 balls numbered 0; 2 balls numbered 3, 3 balls numbered 4, and 1 ball numbered 6;
2. The urn has 8 balls numbered 0; 1 ball numbered 5, and 1 ball numbered 10.

If you get to choose with which urn to play, what would be your preference? Why?

Exercise 2.34 Suppose Jerry inspects a soda company and determines that about 1 out of every 200 plastic bottles coming out of the production line has a choked neck defect. He selects three bottles at random. What is the expected number of those with a choked neck defect?

Exercise 2.35 You draw two cards from a standard deck of 52 cards. What is the expected number of aces you will draw?

2.8 Some Common Random Variables

For each example in the previous section, we defined a random variable, calculated its probability distribution, and discussed its expected value. The workload may be reduced considerably in some cases, as there are some probability distributions that occur commonly in a broad range of practical problems. It is convenient to identify conditions under which they arise and study the general case. Then, when we can determine that a random variable in a specific problem satisfies those conditions, we can apply the general results and solve our problem more efficiently.

In what follows, we present a few such random variables. We give the framework within which they occur, find their probability distribution, and calculate their expected values.

2.8.1 The Bernoulli Distribution

The Bernoulli distribution describes a random variable with only two possible outcomes. It is also sometimes called a binary random variable. In a generic context, the two possible outcomes are usually referred to as "success" and "failure" and represented numerically as 1 and 0, respectively. An experiment with only two possible outcomes is often called a *Bernoulli trial.* In applications, one usually begins by defining what "success" means.

■ **Example 2.26** You play the following game: roll a die once. If you roll a 2 or a 5, you win. If you roll anything else, you lose. Let X be the random variable indicating whether you won or lost. It is natural to call the win a "success" and the loss – a "failure." The probability for success is $\frac{2}{6} = \frac{1}{3}$ and the probability for a loss is $\frac{4}{6} = \frac{2}{3}$. Thus, the distribution of X is given in Table 2.3.

x	$P(X = x)$
1	$\frac{1}{3}$
0	$\frac{2}{3}$

Table 2.3: The distribution of the Bernoulli random variable from Example 2.26.

■

Note that when a die is rolled, there are six possible outcomes, so the outcome of this experiment is not binary. We have $\Omega = \{1,2,3,4,5,6\}$. However, having defined "success"= $\{2, 5\}$ and "failure" = $\{1,3,4,6\}$, we now have a binary experiment (a Bernoulli trial), which we use to define the random variable X with

$$P(\text{success}) = P(X = 1) = \frac{1}{3}, \text{ and } P(\text{failure}) = P(X = 0) = \frac{2}{3}.$$

Definition 2.8.1 Let p be a number with $0 < p < 1$. We say that a binary random variable X is a *Bernoulli random variable* with probability for success p, if

$$P(\text{success}) = P(X = 1) = p, \text{ and } P(\text{failure}) = P(X = 0) = 1 - p.$$

The distribution presented in Table 2.4 is called the *Bernoulli distribution*.

The expected value of a Bernoulli random variable is

$$\mathbb{E}(X) = 1 \cdot p + 0 \cdot (1 - p) = p.$$

x	$P(X = x)$
1	p
0	$1 - p$

Table 2.4: The Bernoulli distribution with probability for success p.

2.8.2 The Geometric Distribution

The framework here is that a Bernoulli trial with a probability for success p is repeated until achieving the first success.

■ **Example 2.27** We revisit Example 2.19. In the game of Trouble, you must roll a 6 on a dice before you can move around the board.

Define "success" to mean you roll a 6 and "failure" to mean you roll 1, 2, 3, 4, or 5. Now the random variable defined in Example 2.19 can be described as the roll on which you get your first success.

The probability for success on the first roll is $\frac{1}{6}$; i.e.,

$$P(X = 1) = \frac{1}{6}.$$

To have $X = 2$, we must have a failure on the first roll and a success on the second. Thus,

$$P(X = 2) = \left(\frac{5}{6}\right) \cdot \left(\frac{1}{6}\right).$$

To have $X = 3$, the first and the second rolls must be failures and the third roll – a success. Thus,

$$P(X = 3) = \left(\frac{5}{6}\right)^2 \cdot \left(\frac{1}{6}\right).$$

Similarly,

$$P(X = 4) = \left(\frac{5}{6}\right)^3 \cdot \left(\frac{1}{6}\right),$$

$$\vdots$$

$$P(X = n) = \left(\frac{5}{6}\right)^{n-1} \cdot \left(\frac{1}{6}\right)$$

$$\vdots$$

and so on. We don't know on which roll we will get the first success and, as noted in Example 2.19, X is a random variable that can take any integer value $n = 1, 2, 3, \ldots$. ∎

The last example gives us a special case of the so-called Geometric distribution.

Definition 2.8.2 Consider a Bernouli trial (binary experiment) with a probability for success p. Repeat that same experiment, independent from previous outcomes, until the first success is achieved. The random variable X that gives the successive number of the repetition on which the first success occurs is a *geometric random variable* with probability for success p. Its distribution is given by

$$P(X = n) = (1-p)^{n-1}p = q^{n-1}p \ \text{ for } n = 1, 2, 3, \ldots, \text{ where } q = 1 - p.$$
$$(2.10)$$

First, we have to check that the probabilities from Equation (2.10) give a distribution; that is, we need to show that their sum is 1. This fact is not obvious, but, in Section 3.5, you will learn how to interpret infinite sums and that

$$1 + r + r^2 + r^3 + \cdots = \sum_{n=0}^{\infty} r^n = \frac{1}{1-r}, \text{ for } r < 1.$$

The infinite sum on the left is called a *geometric series*. In our case,

$$p + qp + q^2 p + q^3 p + \cdots = p \sum_{n=0}^{\infty} q^n = \frac{p}{1-q} = \frac{p}{p} = 1,$$

since $q = 1 - p$. Therefore, the probabilities from Equation (2.10) provide a probability distribution.

To compute the expected value for this geometric distribution, we have

$$E(X) = \sum_{n=1}^{\infty} nP(X = n) = \sum_{n=1}^{\infty} nq^{n-1} p.$$

Evaluating this infinite sum can be done when we notice that it is related to a geometric series.

We know

$$1 + q + q^2 + q^3 + \cdots = \frac{1}{1-q}.$$

Differentiating both sides, we have

$$\frac{d}{dq}(1 + q + q^2 + q^3 + \cdots) = \frac{d}{dq}\left(\frac{1}{1-q}\right),$$

which gives us

$$1 + 2q + 3q^2 + \cdots = \sum_{n=1}^{\infty} nq^{n-1} = \frac{1}{(1-q)^2}.$$

So,

$$\mathbb{E}(X) = \sum_{n=1}^{\infty} nq^{n-1} p = p \sum_{n=1}^{\infty} nq^{n-1} = \frac{p}{(1-q)^2} = \frac{1}{p}. \tag{2.11}$$

■ **Example 2.28** Let's now return to Example 2.27, where X is the number of the roll on which you got a 6 for the first time. Now, X is a geometric random variable with $p = \frac{1}{6}$ and $q = \frac{5}{6}$. Using the result we just obtained for the geometric distribution, gives us $\mathbb{E}(x) = \frac{1}{p} = 6$. Thus, in a game of Trouble, we would expect that, on average, you will have to roll six times before getting the first six. This should make intuitive sense. ■

There is another random variable that we can consider, which is closely related to the geometric distribution, as given in Definition 2.10. Instead of considering the repetition number X at which the first success is achieved, we can define a random variable Y, the values of which give *the number of failures prior to the first success*. Thus, if the first experiment is a success, we will have $Y = 0$; if the sixth experiment is the first success, we know that there have been 5 prior failures, so $Y = 5$, and so on. In general $Y = X - 1$. Thus Y can take values $\{0, 1, 2, 3, \dots\}$.

The distribution of Y is given by

$$P(Y = n - 1) = (1 - p)^{n-1}p = q^{n-1}p \quad \text{for } n = 1, 2, 3, \dots. \qquad (2.12)$$

Very often, this random variable is also called geometric, so we have to be clear which of the two versions an applied problem uses.

Exercise 2.36 An unfair coin has a probability of 0.6 of getting heads. What is the probability the first tails occurs on the fourth flip?

Exercise 2.37 You have a large lot of auto parts 80% of which are good and 20% defective. You draw 1 at a time.
1. What is the probability that you get the first defective on the fourth draw?
2. On average, how many parts do you expect to have to draw until you hit the first defective?

Exercise 2.38 We roll an 8-sided die. What is the expected number of rolls until a 3 appears?

Exercise 2.39 In a certain hospital, long-term records show that 40% of the newborns are male.
1. On a given day, what is the probability the first female is the third baby born for the day?
2. What is the expected number number of births before the first girl for the day is born?

Exercise 2.40 Suppose Jerry inspects a soda company and determines that about 1 out of every 200 plastic bottles coming out of the production line has a choked neck defect.

1. Jerry wants to know the likelihood that the first defective bottle he finds is the fifth one that he tests.
2. Jerry wants to know the likelihood that it takes at most six trials until he finds the first defective bottle.
3. Jerry wants to know the likelihood that it takes at least ten trials until he finds the first defective bottle.

Exercise 2.41 Show that for the alternative form of a geometric random variable with the distribution given in Equation (2.12), the expected value is

$$E(Y) = \frac{1-p}{p}.$$

Hint. Show that

$$E(Y) = \sum_{n=1}^{\infty} (n-1)(1-p)^{n-1}p = E(X) - 1 = \frac{1-p}{p}.$$

2.8.3 The Binomial Distribution

This distribution arises when a positive integer n is selected, and then we perform n independent Bernoulli trials with probability for success p. We consider the random variable X that gives the number of successes in the sequence of those n trials.

■ **Example 2.29** Let's choose $n = 4$, which means we will have to repeat our binary experiment 4 times. For each trial, we record if it was a success (S) or a failure (F). Let's say we obtained

    $FSSS$.

Because there were three successes, in this case the value of the random variable is $X = 3$.

We now want to find the distribution of X. First, notice that X may take values $0, 1, 2, 3, 4$. There is only one possible outcome for which $X = 0$: $FFFF$. Similarly, $X = 4$ only for the case $SSSS$. When X has a value $1, 2$, or 3, there are more possibilities which are listed in the second column of Table 2.5.

Each sequence listed in the second column of Table 2.5 is an elementary event in the sample space Ω comprised of all sequences of length 4 that can be

x	Outcomes for which $X = x$	$P(X = x)$
0	$FFFF$	$(1-p)^4$
1	$SFFF,\ FSFF,\ FFSF,\ FFFS$	$4(1-p)^3 p$
2	$SSFF,\ SFSF,\ SFFS,\ FSSF,\ FSFS,\ FFSS$	$6(1-p)^2 p^2$
3	$SSSF,\ SFSS,\ SSFS,\ FSSS$	$4(1-p)p^3$
4	$SSSS$	p^4

Table 2.5: The distribution of X, the number of successes, for $n = 4$ and probability for success p.

formed from the letters F and S. The probability of each elementary event is calculated as a product of length 4, where $(1-p)$ is the probability for failure and p is the probability of success. For example,

$$P(SFSS) = p \cdot (1-p) \cdot p \cdot p = (1-p)p^3.$$

Then

$$P(X = 3) = P(SSSF) + P(SFSS) + P(SSFS) + P(FSSS) = 4(1-p)p^3,$$

since the probability of each sequence is $(1-p)p^3$. ∎

In this example, we were able to list all possible elementary events that give us $P(X = x)$ for $x = 0, 1, 2, 3, 4$. For larger values of n, this will not be practical. However, this example helps us as a stepping stone toward the general case.

Note that each row in Table 2.5 lists all sequences that have x number of S's, where the value of x is given in the first column. The probability of each such sequence is $(1-p)^{4-x}p^x$. Thus, to calculate $P(X = x)$, we simply need to know how many such sequences there are. We approach this question by asking in how many ways we can select x number of places in a sequence of length 4 to put the S's. This is a counting problem, to which we already know the answer! According to Theorem 2.3.3, we can choose x places from the 4 available in this many ways:

$$\binom{4}{x} = \frac{4!}{x!(4-x)!}, \quad \text{for } x = 0, 1, 2, 3, 4.$$

Now we obtain that

$$P(X = x) = \binom{4}{x}(1-p)^{4-x}p^x. \tag{2.13}$$

The definition below is a straightforward generalization for an arbitrary number of trials n.

Definition 2.8.3 Let n be a positive integer. Assume that n independent Bernoulli trials are performed, each with probability for success p. Let X be the random variable that gives *the number of successes* among the n trials. Then

$$P(X = x) = \binom{n}{x}(1-p)^{n-x}p^x \text{ for } x = 0,1,2,3,\ldots,n. \qquad (2.14)$$

We say that the random variable X has a *Binomial distribution* with parameters n and p.

■ **Example 2.30** You roll a die 10 times. Each time you roll a 1, you win $1. What is the probability that after the ten rolls you have won $6.

Here, if we define "success" to mean "we roll a 1," we consider 10 Bernoulli trials with probability for success $p = \frac{1}{6}$. If X is the number of successes among the ten trials, X has a binomial distribution with parameters $n = 10$ and $p = \frac{1}{6}$. From Equation (2.14) we calculate

$$P(X = 6) = \binom{10}{6} \cdot \left(\frac{5}{6}\right)^{10-6}\left(\frac{1}{6}\right)^6 = \frac{10!}{4!6!}\left(\frac{5}{6}\right)^4\left(\frac{1}{6}\right)^6 \approx 0.0022.$$

■

Using the Binomial Theorem from Section 6.5 of the Appendix we can see that Equations 2.14, for $x = 1,2,3,\ldots,n$, indeed define a probability distribution because

$$1 = 1^n = \left(p + (1-p)\right)^n = \sum_{x=0}^{n}(1-p)^{n-x}p^x.$$

It can be shown that if X is a binomial random variable with parameters n and p,

$$\mathbb{E}(x) = np(1-p).$$

The proof is not difficult but is beyond the scope of these notes. You can find it, e.g., in [1].

Exercise 2.42 You have a large lot of auto parts 80% of which are good and 20% defective. You draw 10 of them.

1. What is the probability that you get only 1 defective part among the 10 drawn?
2. What is the probability that you will get 6 defectives among the 10 drawn?
3. If you repeat the process of drawing 10 auto parts from the lot many, many times, what would you expect for the average number of defective parts among the 10 be?

Exercise 2.43 Suppose Jerry inspects a soda company and determines that 1 out of every 200 plastic bottles coming out of the production line has a choked neck defect. He selects three bottles at random from the production line. What is the probability that exactly one will have a chocked neck defect?

Exercise 2.44 In a certain hospital, long-term records show that 40% of the newborns are male. The records show that there were 8 births recorded in that hospitals on October 7, 2023.

1. What is the probability that exactly 6 were boys?
2. What is the probability that at least 6 of the babies born on that day were girls?

Exercise 2.45 You pay $10 to play the following game: Draw 5 cards from a standard deck of 52 cards with replacement. (After each card drawn, you put the card back in the deck and reshuffle well before drawing again.) For each spade drawn, you win $4. What is the probability that your net win will be at least $2?

Exercise 2.46 A machine produces an item and we know that 1 out of 15 produced is defective. In a sample of 8 items, what is the probability that exactly 3 are defective?

Exercise 2.47 A fair coin is flipped six times. What is the probability that exactly 3 are heads?; exactly 4 are heads?

Exercise 2.48 Verify that when using Equation (2.13) to find $P(X = x)$, we obtain the probabilities given in Table 2.5.

2.9 Summary and What to Expect Next

The field of probability arose from gamblers wanting to know their odds of winning, but nowadays executives often base their decisions on the likelihood that something will or will not occur. Thus, understanding the rules that control randomness and how to apply them to increase the likelihood of success are important for addressing essential questions in biology, economics, medicine, ecology, business, social science, finance, computing, and much more. To understand probability is to understand that randomness and variability are natural, predictable, and quantifiable. The mathematical theory of probability gives us a road map. In this chapter, we introduced you to some of the terminology and basic principles of random events with discrete number of outcomes, e.g., experiments or random variables with only finite or countable number of outcomes.

When you take a more advanced probability course, you will also examine random events and random variables that can produce a continuous range of outcomes. This will include continuous random variables, a certain class of non-decreasing real-valued functions that define continuous distributions, and important limit theorems that link some of those distributions. In that general context, the mathematical theory requires the use of Calculus, where probabilities are determined by integrating appropriate functions, called probability density functions. At an even more advanced level (that is, at the level of graduate courses), probability theory also connects with functional analysis and measure theory. Having now worked through this chapter, you have a head start on your traditional calculus-based probability course. And, in case the material has captured your interest, we have also opened a door for you to the broader theory of probability that can lead you much further into this field.

2.10 Suggested Further Reading

We recommend the following textbooks as further reading. They are standard texts used for undergraduate courses in Probability.

Bibliography

[1] Scheaffer, Richard L., and Linda J. Young. *Introduction to Probability and Its Applications*, 3rd Ed. Brooks Cole, Cengage Learning, 2010.

[2] Grinstead, Charles, and Laurie J. Snell. *Introduction to Probability*. American Mathematical Society, 2006.

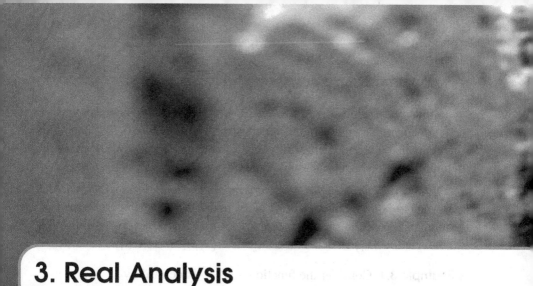

3. Real Analysis

.1 Introduction

Real Analysis is a broad area of mathematics devoted to the rigorous study of the set of real numbers and their properties, infinite sequences and series of real numbers, and real functions. Most of you already have a head start on the basic topics and are familiar with some of the terminology, which is the same as that used in calculus. In a Real Analysis course, however, the focus is to provide rigorous mathematical definitions and proofs for objects and properties treated mainly as computational tools in a traditional Calculus course. Real analysis is a vast field of study, and an introductory course can usually only cover the fundamentals. To emphasize this, the course is also offered at some institutions under the name Advanced Calculus.

Results from real analysis are used to solve problems in a variety of disciplines, including physics, engineering, economics, and biology. Real analysis also provides a sound theoretical foundation for other areas of mathematics such as differential equations, advanced probability theory, statistics, and functional analysis.

.2 Sequences of Real Numbers

In this section, we focus on a topic that you have already seen in your Calculus courses. Those courses have helped you develop intuitive understanding of how sequences of real numbers behave, whether they converge, and how to

DOI: 10.1201/9781032623849-3

determine what values they converge to. In a Real Analysis course, you will
revisit the material and learn how to construct rigorous proofs for many of the
theorems listed without proof in your Calculus textbooks. We begin with a
definition, with which you are likely familiar.

Definition 3.2.1 A *sequence of real numbers* (hereafter, merely called a
sequence) is a function $f : \mathbb{N} \to \mathbb{R}$, from the positive integers to the real
numbers. We often write $x_n = f(n)$, for $n = 1, 2, 3, \ldots$. We will use the
abbreviation $\{x_n\}$ to denote the sequence $\{x_1, x_2, x_3.x_4, \ldots\}$.

■ **Example 3.1** Consider the function $f : \mathbb{N} \to \mathbb{R}$ defined by

$$x_n = f(n) = 3n + 5, \text{ for } n = 1, 2, 3, \ldots.$$

Now, $x_1 = f(1) = 8$, $x_2 = f(2) = 11$, $x_3 = f(3) = 14$, and so on, giving the
terms of the sequence $\{x_n\}$. We should keep in mind that this is an ordered
list; that is, *the order in which we list the values is important.*   ■

Two sequences are equal if and only if they are equal term by term. So the
sequence $\{1, 2, 3, 4, 5, 6, \ldots\}$ is not equal to the sequence $\{2, 1, 3, 4, 5, 6, \ldots\}$,
even though they may be the same numbers.

It is common practice to switch between listing the elements of a sequence
and the function defining the sequence. For example,
$\{\frac{1}{2}, \frac{1}{4}, \frac{1}{8}, \frac{1}{16}, \ldots\}$ could be described by $\{\frac{1}{2^n}\} = \{x_n\}$. We have to be careful
here, though, as the latter is valid only when we are certain that the pattern
identified by the first few elements of the sequence holds for all of its terms.

If one were to graph a sequence, it would be a graph such as the one shown
in Figure 3.1.

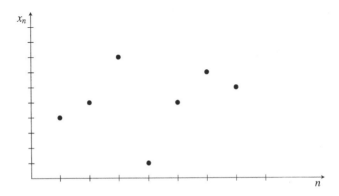

Figure 3.1: Graph of a sequence with terms $\{4, 5, 8, 1, 5, 7, 6, \ldots\}$.

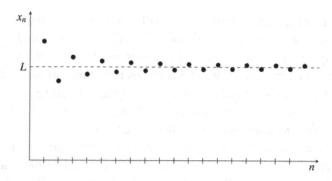

Figure 3.2: A plot of a sequence that converges to the number L.

The question of most interest to us in this section will be to decide if a sequence converges. Intuitively, a sequence $\{x_n\}$ converges if, when we go far out in the list of numbers in the sequence, the numbers get arbitrarily close to some fixed value (see Figure 3.2).

It took approximately 200 years after the invention of calculus to arrive at the rigorous definition of convergence that we use today, which we now state.

Definition 3.2.2 The sequence $\{x_n\}$ *converges to the number L* if, for any number $\varepsilon > 0$, there is a number $N(\varepsilon)$, so that if $n > N(\varepsilon)$, then $|x_n - L| < \varepsilon$.

When a sequence $\{x_n\}$ converges to L, we write $x_n \to L$ or $\lim x_n = L$.

A sequence that doesn't converge is said to diverge and is called a *divergent sequence.*

Many consider this definition among the most important in mathematics, and it is perhaps the most important definition in real analysis. Let's unpack what it means geometrically. Recall (see Section 6.2 in the Appendix) that $|x_n - L|$ refers to the distance of x_n on the number line to the number L. Thus $|x_n - L| < \varepsilon$ states that the distance from x_n to L is smaller than ε. That is, the number $x_n - L$ (no absolute values!) satisfies

$$-\varepsilon < x_n - L < \varepsilon,$$

or, equivalently,

$$L - \varepsilon < x_n < L + \varepsilon.$$

Graphically, this tells us that x_n is between the horizontal lines labeled $L - \varepsilon$ and $L + \varepsilon$ in Figure 3.3. Now, if the sequence converges to L, the definition requires that no matter how small $\varepsilon > 0$ is, we can find a cutoff $N = N(\varepsilon)$, so that for indices $n > N$, *all* values x_n will be in the band $(L - \varepsilon, L + \varepsilon)$.

We say that the cutoff value N depends on ε, and write $N = N(\varepsilon)$ to emphasize this dependence. What this means is that the cutoff N will change depending on the chosen value of $\varepsilon > 0$. Looking at Figure 3.3, the distance between the horizontal lines labeled $L - \varepsilon$ and $L + \varepsilon$ will widen for larger values of ε and get narrower for smaller values of ε.

If we take a smaller value of ε than the one chosen to draw Figure 3.3, that will squeeze the band $(L - \varepsilon, L + \varepsilon)$ about L. Then some of the values in the figure that are now within the band will no longer be in it. Then, the number N will have to move to the right. The important point is that with that new N, *all x_n with $n > N$ will again be in the band $(L - \varepsilon, L + \varepsilon)$.*

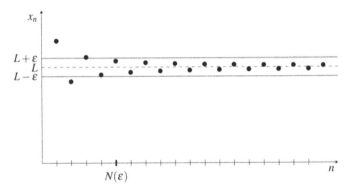

Figure 3.3: Graphical meaning of Definition 3.2.2. For the chosen ε, we have found a number $N(\varepsilon)$, such that for all $n > N(\varepsilon)$, the values x_n of the sequence are between the lines $y = L - \varepsilon$ and $y = L + \varepsilon$.

An alternative depiction is also quite helpful for understanding the definition of convergence. Suppose the sequence $\{x_n\}$ converges to L. We draw a number line and locate the number L. Now draw an interval around L with a small radius ε. The definition says that no matter how small $\varepsilon > 0$ is, all terms of the sequence after N are in this interval (see Figure 3.4). This means that if $x_n \to L$, all elements of $\{x_n\}$ *except, possibly, for finitely many* (namely, $x_1, x_2, \ldots, x_N$) are in the interval $(L - \varepsilon, L + \varepsilon)$.

As a shorthand, when we know that a sequence $\{x_n\}$ converges to L, we will often say that we can make $|x_n - L|$ arbitrarily small for all *sufficiently large n*.

When you use Definition 3.2.2 to show that a sequence $\{x_n\}$ converges, you will need to have a guess for the value L in mind. We usually make that guess by either graphing the sequence $\{x_n\} = f(n)$ to see if the values appear

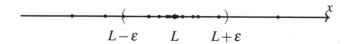

Figure 3.4: Pick any $\varepsilon > 0$. If $\{x_n\}$ converges to L, Definition 3.2.2 says that all terms of the sequence after x_N are in the interval $(L - \varepsilon, L + \varepsilon)$.

to approach a value L, or use other techniques that you have learned in your Calculus course. The important thing is that such a value L is just a guess until you prove $\{x_n\} \to L$, using Definition 3.2.2.

Our first example demonstrates how to use the definition to prove that a sequence converges.

■ **Example 3.2** Prove that $\{x_n\} = \{\frac{1}{n}\}$ converges to $L = 0$.

Intuitively, this should again make sense – as n gets large, the fraction $\frac{1}{n}$ gets close to 0. We begin the proof by imagining of our worst enemy picking the smallest positive number they can think of. (Of course, there is no smallest positive number.) So, the first sentence of the proof should be "Let $\varepsilon > 0$ be given." We have to convince a skeptic that no matter how small the chosen number ε is, we can find $N(\varepsilon)$ so that, if $n > N(\varepsilon)$, then

$$\left| \frac{1}{n} - 0 \right| = \left| \frac{1}{n} \right| = \frac{1}{n} < \varepsilon.$$

The last inequality is the same as $n > \frac{1}{\varepsilon}$. Thus, if we choose $N(\varepsilon) = \frac{1}{\varepsilon}$, we will have $|\frac{1}{n} - 0| < \varepsilon$ for all integers $n > N(\varepsilon)$. Thus, by Definition 3.2.2, the sequence $\{\frac{1}{n}\}$ converges to 0.

Notice that $N(\varepsilon)$ depends on ε; in fact, no number N can work for every ε. That is, first ε is fixed, and then we find the N that depends on the choice of ε.   ■

■ **Example 3.3** Prove that the sequence $\{x_n\} = \{1 + \frac{1}{2n}\}$ converges to $L = 1$.

Intuitively, this should again make sense: as n becomes large, $\frac{1}{2n}$ becomes closer to 0, so the expression

$$1 + \frac{1}{2n} \quad \text{should be getting closer to } L = 1.$$

To prove this using the definition, we again go through the following steps:
1. We chose an $\varepsilon > 0$ and fix it.
2. We find how large n must be to ensure that

$$|a_n - L| = \left| \left(1 + \frac{1}{2n} \right) - 1 \right| < \varepsilon.$$

Now

$$|a_n - L| = \left|\left(1 + \frac{1}{2n}\right) - 1\right| = \frac{1}{2n}.$$

To have $\frac{1}{2n} < \varepsilon$, we solve for n to get that

$$n > \frac{1}{2\varepsilon}.$$

Thus, if $N(\varepsilon)$ is at least as large as $\frac{1}{2\varepsilon}$, that will ensure that for all positive integers $n > N(\varepsilon)$, we have $|a_n - L| < \varepsilon$.

Notice again that $N(\varepsilon)$ depends on ε. That is, first ε is fixed, and then $N(\varepsilon)$ is found.   ■

■ **Example 3.4** We will show that the sequence $\{\frac{1}{3^n}\}$ converges to 0. The algebraic manipulations here are a bit more complex, but the idea is the same as in the previous examples.

Let $\varepsilon > 0$ be given. We want to show that we can find $N(\varepsilon)$, so that if $n > N(\varepsilon)$, then

$$\left|\frac{1}{3^n} - 0\right| = \left|\frac{1}{3^n}\right| = \frac{1}{3^n} < \varepsilon.$$

Finding this n amounts to solving the above inequality for n, so we need to solve

$$\frac{1}{3^n} < \varepsilon \text{ or, equivalently, } \frac{1}{\varepsilon} < 3^n$$

Taking natural logarithms from both sides leads to

$$\ln\frac{1}{\varepsilon} < \ln 3^n = n\ln 3, \text{ so } n > \frac{\ln\frac{1}{\varepsilon}}{\ln 3}.$$

Thus, we take

$$N(\varepsilon) = \frac{\ln\frac{1}{\varepsilon}}{\ln 3}.$$

For example, if $\varepsilon = 10^{-6}$, then $N(10^{-6}) = \frac{\ln 10^6}{\ln 3}$.   ■

Example 3.4 is a special case of the following more general fact.

■ **Example 3.5** Let r be a real number with $|r| < 1$. Then the sequence $\{x_n\} = \{|r^n|\}$ converges to 0.

Let $\varepsilon > 0$. We want to show that we can find $N(\varepsilon)$ so that if $n > N(\varepsilon)$, then

$$||r^n| - 0| = |r^n| < \varepsilon.$$

Finding this n amounts to solving the above inequality for n, so we need to solve

$$|r^n| = |r|^n < \varepsilon.$$

Taking natural logarithms from both sides leads to

$$\ln |r|^n < \ln \varepsilon, \text{ or, equivalently, } n \ln |r| < \ln \varepsilon.$$

Since $|r| < 1$, $\ln |r| < 0$. Thus, dividing the above inequality by $\ln |r|$ shows that we need to have

$$n > \frac{\ln \varepsilon}{\ln |r|}.$$

Thus, we choose

$$N(\varepsilon) = \frac{\ln \varepsilon}{\ln |r|}.$$

This guarantees that for all integers $n > N(\varepsilon)$, we have $||r^n| - 0| < \varepsilon$. Therefore, we have proved that

$$\lim |r^n| = 0, \text{ for } |r| < 1.$$

■

The next example shows how we can prove that a sequence diverges.

■ **Example 3.6** Consider the sequence $\{1, -1, 1, -1, 1, -1, \ldots\}$; that is, the sequence $x_n = (-1)^n$. We will prove the sequence diverges, using a proof by contradiction (see Section 1.3.4).

Assume the sequence converges to a limit L. Now let $\varepsilon > 0$ be given. Because the sequence converges, there is an N, so that all x_n for $n > N$ are in the interval $(L - \varepsilon, L + \varepsilon)$; that is, only the first $n - 1$ terms may be outside of this interval. But if we consider $\varepsilon = \frac{1}{2}$, at least one of 1 and -1 lies outside of $(L - \frac{1}{2}, L + \frac{1}{2})$, regardless of what L is. To see this, note that the distance between -1 and 1 on the number line is 2, and the length of the interval $(L - \frac{1}{2}, L + \frac{1}{2})$ is 1, so it cannot cover both -1 and 1. Thus the sequence has infinitely many terms outside of $(L - \frac{1}{2}, L + \frac{1}{2})$, which contradicts the assumption that the sequence converges. Therefore $\{(-1)^n\}$ diverges. ∎

■ **Example 3.7** Consider the sequence

$$\{a_n\} = \{2, 4, 6, 8, 10, \dots\},$$

with $a_n = 2n$. Prove that the sequence diverges.

Again, we will use a proof by contradiction. We start by assuming (contrary to what we need to prove) that $\{a_n\}$ does converge to a limit L. Now let's choose $\varepsilon = \frac{1}{2}$. Since we assumed $\{a_n\}$ converges to L, there is an $N = N(\varepsilon)$ such that

$$|a_n - L| < \frac{1}{2}.$$

This means (see again Section 6.2 in the Appendix) that

$$a_n \in (L - \frac{1}{2},\ L + \frac{1}{2}) \text{ for } all\ n > N.$$

This is a contradiction, since the terms of the sequence $a_n = 2n$ get arbitrarily large for large n. Therefore, the sequence $\{a_n\}$ diverges. ∎

In the last example, we used the phrase "arbitrarily large." Intuitively, it should make sense, but how can we define it more rigorously? What we want to say when we use that phrase is that no matter which very large number α we pick, we can always find an element a_n of the sequence such that $a_n > \alpha$.

Example 3.7 illustrates a special type of divergence.

Definition 3.2.3 We say that a sequence $\{a_n\}$ *diverges to infinity* (diverges to ∞) when for any number α, *all* elements of the sequence, except (possibly) for finitely many, satisfy $a_n > \alpha$. More specifically, $\{a_n\}$ diverges to ∞ if for any number α, there is a number $N = N(\alpha)$, such that for all integers $n > N$, $a_n > \alpha$.

We say that a sequence $\{a_n\}$ *diverges to negative infinity* (diverges to $-\infty$) if for any number β, all elements of the sequence, except (possibly) for finitely many, satisfy $a_n < \beta$. More specifically, $\{a_n\}$ diverges to $-\infty$ if for any number β, there is a number $N = N(\beta)$, such that for all integers $n > N$, $a_n < \beta$.

■ **Example 3.8** Consider the sequence

$$\{x_n\} = \{1, 2, 1, 4, 1, 6, 1, 8, 1, 10, \dots\}.$$

Show that the sequence diverges but that it does not diverge to infinity.

The proof that $\{x_n\}$ diverges is the same as in Exercise 3.7. In this case, however, x_n does not diverge to ∞. To see this, take any $\alpha > 1$ (e.g., $\alpha = 2$). As any other element of $\{x_n\}$ is equal to 1, there are now *infinitely many* elements of the sequence that satisfy $x_n < \alpha$, in contrast with the requirements in Definition 3.2.3. Thus, $\{x_n\}$ does not diverge to ∞.   ■

> **Theorem 3.2.1** A sequence $\{x_n\}$ cannot converge to more than one number.

Proof. This theorem is proved by contradiction. Since we want to prove that if $\{x_n\}$ converges, the limit value is unique, we will suppose the contrary. That is, suppose that $\{x_n\}$ converges to two different numbers L and M.

Since $M \neq L$, we can take

$$\varepsilon = \frac{|L - M|}{4} > 0,$$

and now the intervals $(M - \varepsilon, M + \varepsilon)$ and $(L - \varepsilon, L + \varepsilon)$ will not overlap (see Figure 3.5). Since $\{x_n\}$ converges to L, there is a number N_1, such that for all $n > N_1$, the terms x_n are in the interval $(L - \varepsilon, L + \varepsilon)$. Since we assumed that $\{x_n\}$ also converges to M, there is a number N_2, such that for all $n > N_2$, the terms x_n are in the interval $(M - \varepsilon, M + \varepsilon)$. Suppose N is larger than both N_1 and N_2. Then all x_n with $n > N$ must be in both intervals. This is impossible, since the intervals do not overlap. We have reached a contradiction. Therefore the limit of a convergent sequence is unique.   ■

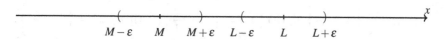

Figure 3.5: If $M \neq L$, and $\varepsilon = \frac{|L-M|}{4} > 0$, the intervals $(M - \varepsilon, M + \varepsilon)$ and $(L - \varepsilon, L + \varepsilon)$ don't overlap.

> **Exercise 3.1** Answer the questions below to further test your understanding of Definition 3.2.2.
>   1. Do you think of ε as a large number or a small number?
>   2. Do you think of $N(\varepsilon)$ as a large number or a small number?

3. Do you choose ε first and then find $N(\varepsilon)$ or vice versa?
4. In a given problem, if ε gets smaller, how do you think $N(\varepsilon)$ will usually change?

Exercise 3.2 Let c be a constant. Prove that the "constant" sequence $\{x_n\}$, $x_n = c$, converges to the constant c.

Exercise 3.3 Prove that the sequence $\{a_n\}$, $a_n = m(-1)^n$ diverges for any real number $m \neq 0$.

Exercise 3.4 Prove that $\{\frac{1}{n^2}\}$ converges to 0. *Hint:* Follow the same steps as in Example 3.2

Exercise 3.5 Let $k > 0$ be an integer. Prove that $\{\frac{1}{n^k}\}$ converges to 0.

Exercise 3.6 Show that the sequence $\{a_n\} = \{-2, -4, -8, -16, -32, \ldots\}$ with $a_n = -2^n$ diverges to $-\infty$.

Exercise 3.7 Show that the sequence $\{x_n\} = \{-2, 4, -8, 16, -32, \ldots\}$ with $x_n = (-1)^n 2^n$ diverges but that it does not diverge to either ∞ or $-\infty$.

Exercise 3.8 Consider the sequences

(i) $\{a_n\} = \left\{3 - \dfrac{1}{4n}\right\}$,   (ii) $\{a_n\} = \left\{\dfrac{2n-3}{3n-1}\right\}$,   (iii) $\{a_n\} = \left\{1 + \dfrac{1}{2^n}\right\}$.

For each of the sequences:
1. Find the number L to which the sequence converges;
2. Find how large n must be, so that $|a_n - L| < 0.1$;
3. If $\varepsilon > 0$, find $N(\varepsilon)$ such that for all $n > N(\varepsilon)$, $|a_n - L| < \varepsilon$.

3.3 Bounds for Sets of Real Numbers

There are some fundamental properties of sequences that we want to prove next. This section introduces some important results and terminology.

> **Definition 3.3.1** A set A of real numbers is *bounded above* if there is a number α for which $x \leq \alpha$ for all $x \in A$. A number α with this property is called an *upper bound* for the set A.

> **Definition 3.3.2** A set A of real numbers is *bounded below* if there is a number β for which $\beta \leq x$ for all $x \in A$. A number β with this property is called a *lower bound* for the set A.

> **Definition 3.3.3** A set A of real numbers is *bounded*, if it is bounded both above and below.

Note that if α is an upper bound for a set of real numbers A, then any number $\alpha_1 > \alpha$ is also an upper bound for A. If β is a lower bound for a set of real numbers A, then any number $\beta_1 < \beta$ is also a lower bound for A.

■ **Example 3.9** The set of numbers

$$A = \{x \in \mathbb{R} \mid x = \frac{1}{n}, \text{ where } x \text{ is a positive integer}\}$$

is bounded above by 1 and below by 0. Thus, this is a bounded set.

The set of numbers

$$\{x \in \mathbb{R} \mid x = 3^n, \text{ where } n \text{ is a positive integer}\}$$

is bounded below by 3, but has no upper bound.

The set of numbers

$$\{x \in \mathbb{R} \mid x = -2n + 5, \text{ where } n \text{ is a positive integer}\}$$

is bounded above by 5, but has no lower bound. ■

These examples show that infinite sets may or may not be bounded. Finite sets, on the other hand, are always bounded.

Theorem 3.3.1 Any finite set A of real numbers is bounded.

Proof. Let A be a set of n elements: $A = \{x_1, x_2, \ldots, x_n\}$, where n is a positive integer. Let α be the largest among those numbers and β be the smallest. Then, for every n,

$$\beta \leq x_n \leq \alpha,$$

proving that the set A is bounded. ■

Definition 3.3.4 We say that a sequence $\{x_n\}$ is bounded if the set of its values is bounded.

■ **Example 3.10** The sequence $\{x_n\} = \{\frac{1}{n}\}$ is bounded, because in Example 3.9 we saw that the set

$$A = \{x \in \mathbb{R} \mid x = \frac{1}{n}, \text{ where } x \text{ is a positive integer}\}$$

is bounded.   ■

Note that if a sequence is not bounded, this means we won't be able to find both an upper bound and a lower bound. Intuitively, this means that the sequence has arbitrarily large or arbitrarily small elements. In other words, if a sequence $\{a_n\}$ is *not* bounded from *above*, no matter which number α we pick, α will not be an upper bound; that is, for any α, we will be able to find an n, for which $a_n > \alpha$. If $\{a_n\}$ is *not* bounded from *below*, no matter which number β we pick, β will not be a lower bound; that is, for any β, we will be able to find an n, for which $a_n < \beta$.

Convergent sequences have the following very important property.

> Theorem 3.3.2 Every convergent sequence is bounded.

Proof. Let $\{x_n\}$ be a convergent sequence and $\lim x_n = L$. By Definition 3.2.2, if we take $\varepsilon = 1$, there is a number N, such that for all $n > N$,

$$|x_n - L| < 1, \text{ which is equivalent to } L - 1 < x_n < L + 1.$$

Thus, the only terms that may not satisfy these inequalities are $x_1, x_2, \ldots, x_N$. Take the set $\{x_1, x_2, \ldots, x_N, L - 1, L + 1\}$. This set is finite and, thus, bounded (Theorem 3.3.1). The bounds on this set give an upper and a lower bound for $\{x_n\}$. So, $\{x_n\}$ is bounded.   ■

The next result shows that information about upper and lower bounds of convergent sequences carry over to their limits.

> Theorem 3.3.3 Suppose that a sequence $\{a_n\}$ converges to L, α is an upper bound for $\{a_n\}$, and β is a lower bound for $\{a_n\}$; that is, suppose $\beta \leq a_n \leq \alpha$ for all n. Then, $\beta \leq L \leq \alpha$.

Proof. We will prove the result for upper bounds. The proof for lower bounds, which is done analogously, is left as an exercise (see Exercise 3.13). We will present a proof by contradiction, using an idea similar to that in the proof of Theorem 3.2.1.

Suppose a_n converges to L and $a_n \leq \alpha$ for all n, but assume (contrary to what we want to prove) that $L > \alpha$. Let's choose

$$\varepsilon = \frac{L - \alpha}{2} > 0, \text{ so } \alpha = L - 2\varepsilon. \tag{3.1}$$

Since $\{a_n\}$ converges to L, there exists $N = N(\varepsilon)$ so that for all $n > N$,

$$|a_n - L| < \varepsilon, \text{ or, equivalently, that } L - \varepsilon < a_n < L + \varepsilon. \tag{3.2}$$

Combining Equations (3.1) and (3.2) gives that

$$\alpha = L - 2\varepsilon < L - \varepsilon < a_n, \text{ for all } n > N.$$

This contradicts that $a_n < \alpha$ for all n. Therefore, $L \leq \alpha$. ∎

Our next theorem describes relationships between the limits of sequences that obey certain term-by-term inequalities. These are proved by combining the results from Theorems 3.3.2 and 3.3.3 and ideas similar to those we used to prove them.

> **Theorem 3.3.4** Let $\{a_n\}$, $\{b_n\}$, and $\{c_n\}$ be sequences of real numbers.
>
> 1. Suppose $0 \leq a_n \leq b_n$ for every n and $b_n \to 0$. Then $a_n \to 0$.
>
> 2. Suppose $a_n \leq b_n$ for every n, $\{a_n\}$ converges to L, and $\{b_n\}$ converges to M. Then $L \leq M$.
>
> 3. Suppose $a_n \leq b_n \leq c_n$ for every n, $a_n \to L$ and $c_n \to L$. Then $b_n \to L$. (This result is often called *The Squeezing Theorem*.)

Proof. 1. We know that $\{b_n\} \to 0$. This means, that for any $\varepsilon > 0$, there is an $N = N(\varepsilon)$, such that for all $n > N$, $|b_n - 0| = |b_n| < \varepsilon$. Since $0 \leq a_n \leq b_n$ for all n, we have that $0 \leq |a_n| \leq |b_n|$. This means that for all $n > N$, we will have $|a_n - 0| = |a_n| < \varepsilon$. This proves $\{a_n\} \to 0$.

2. We will use a proof by contradiction. The idea is the same as for the proof of Theorem 3.2.1. Assume that $L > M$ and choose $\varepsilon = \frac{L-M}{4} > 0$. Then, the intervals $(M - \varepsilon, M + \varepsilon)$ and $(L - \varepsilon, L + \varepsilon)$ are disjoint, with $(M - \varepsilon, M + \varepsilon)$ being to the left of $(L - \varepsilon, L + \varepsilon)$ on the number line (see Figure 3.5). Since $\lim a_n = L$, we know that for sufficiently large n, all terms of $\{a_n\}$ will be in the interval $(L - \varepsilon, L + \varepsilon)$. Similarly, since $\lim b_n = M$, all terms of $\{b_n\}$ will be in $(M - \varepsilon, M + \varepsilon)$ for sufficiently large n. This means that for all sufficiently large n, we will have that $b_n < a_n$, which contradicts $a_n \leq b_n$ for all n. Thus, $L \leq M$.

3. Since we know that $a_n \leq b_n \leq c_n$ for all n, part 2 implies that $L = \lim a_n \leq \lim b_n \leq \lim c_n = L$. This shows that $\lim b_n = L$. ∎

We next look at a special class of sequences.

Definition 3.3.5 Let $\{a_n\}$ be a sequence of real numbers.
1. If for every n, $a_n \leq a_{n+1}$, the sequence is *monotone increasing*. If $a_n < a_{n+1}$ for every n, the sequence is said to be *strictly monotone increasing*;
2. If for every n, $a_n \geq a_{n+1}$, the sequence is *monotone decreasing*. If $a_n > a_{n+1}$ for every n, the sequence is said to be *strictly monotone decreasing*.

■ **Example 3.11** The sequence $\{a_n\}$ with $a_n = 0.123\ldots n$ is strictly monotone increasing.

We need to show that $a_n < a_{n+1}$ for all n. Equivalently, we can show that $a_{n+1} - a_n > 0$ for all n. Since we have

$$a_{n+1} - a_n = 0.12\ldots n(n+1) - 0.12\ldots n = 0.\underbrace{00\ldots0}_{n}(n+1) > 0, \text{ for all } n,$$

this proves the sequence is strictly monotone increasing. ■

■ **Example 3.12** The sequence $\{a_n\}$ with $a_n = \frac{1}{2^n}$ is strictly monotone decreasing.

The definition tells us, we need to show that $a_n > a_{n+1}$ for all n. Equivalently, here we can check that $\frac{a_n}{a_{n+1}} > 1$ for all n. In this case, we have

$$\frac{a_n}{a_{n+1}} = \frac{\frac{1}{2^n}}{\frac{1}{2^{n+1}}} = 2 > 1,$$

proving that $\{a_n\}$ is strictly monotone decreasing. ■

■ **Example 3.13** The sequence $\{a_n\}$ defined by $a_n = \sqrt{n+1} + \sqrt{n}$, for $n \geq 1$, is strictly monotone increasing.

We need to show $a_n < a_{n+1}$ for all $n \geq 1$ or, equivalently, that $a_{n+1} - a_n > 0$, for all $n \geq 1$. We have

$$a_{n+1} - a_n = (\sqrt{n+2} + \sqrt{n+1}) - (\sqrt{n+1} + \sqrt{n})$$
$$= \sqrt{n+2} - \sqrt{n} > 0, \text{ for all } n \geq 1.$$

Therefore, $\{a_n\}$ is strictly monotone increasing. ■

Increasing and decreasing sequences have important properties we will study next. Before that, however, we will need to define some additional terms related to upper and lower bounds of sets. Recall that if a set is bounded above (or below), it has infinitely many upper (lower) bounds. Those of most interest are the values that bound the set "most tightly." We give the exact definitions next.

Definition 3.3.6 A number α is the *least upper bound* for a set $A \subseteq \mathbb{R}$, if:
1. α is an upper bound for A, and
2. α is the smallest upper bound; that is, if γ is also an upper bound for A, then $\alpha \leq \gamma$.

Definition 3.3.7 A number β is the *greatest lower bound* for a set $A \subseteq \mathbb{R}$, if:
1. β is a lower bound for A, and
2. β is the largest lower bound; that is, if δ is also a lower bound for A, then $\beta \geq \delta$.

The following theorems give equivalent descriptions for the least upper bound and greatest lower bound that are often helpful in the context of convergent sequences.

Theorem 3.3.5 A number α is the least upper bound for a set of real numbers A if and only if
1. α is an upper bound for A, and
2. given any $\varepsilon > 0$, there is a number $x = x(\varepsilon) \in A$ for which $x(\varepsilon) > \alpha - \varepsilon$.

Proof. First, let's give an intuitive argument. By definition, α is the smallest upper bound. This means that any number α_1 smaller than α cannot be an upper bound. Thus, there will be at least one number in the set A that is larger than α_1.

We now make this intuitive argument rigorous. First, notice that the conditions 1 in both Theorem 3.3.5 and Definition 3.3.6 are the same. Thus, we need to show that conditions 2 are equivalent. Next, notice that this is an *if and only if* proof. This means that we need to provide *two* proofs (see Section 1.3.6). Namely, we need to prove the following claims:

(i) If α the smallest upper bound for A, then condition 2 in Theorem 3.3.5 holds, and

(ii) If condition 2 in Theorem 3.3.5 holds, then α is the smallest upper bound.

Proof of statement (i): We will do a proof by contradiction. Suppose condition 2 of the definition is satisfied, but condition 2 of the Theorem is not. That is, suppose α is the least upper bound, but there is a number ε_0 for which all $x \in A$ will satisfy $x \leq \alpha - \varepsilon_0$. Now, by Definition 3.3.1, this shows $\alpha - \varepsilon_0$ is an upper bound for A. Thus, we have found an upper bound for A, which is smaller than α. This is a contradiction, since we know α is the least upper bound for A.

Proof of statement (ii): Now assume that given any $\varepsilon > 0$, there is a number $x = x(\varepsilon) \in A$ for which $x(\varepsilon) > \alpha - \varepsilon$. This tells us that for any $\varepsilon > 0$, $\alpha - \varepsilon < \alpha$ is not an upper bound. This shows α is the smallest upper bound for A. ∎

Theorem 3.3.6 A number β is the greatest lower bound for a set of real numbers A if and only if
1. β is a lower bound for A, and
2. given any $\varepsilon > 0$, there is a number $x = x(\varepsilon) \in A$ for which $x(\varepsilon) < \beta + \varepsilon$.

Proof. The proof follows the same argument as the proof or Theorem 3.3.5. We leave it as an exercise (see Exercise 3.12). ∎

In Theorem 3.3.2, we proved that a convergent sequence is always bounded. Revisiting Example 3.6 shows that the converse is not true in general: even though the sequence

$$\{x_n\} = \{1, -1, \ 1, -1, \ 1, -1 \ldots\}$$

is bounded, we know that it does not converge. However, our next result shows that the converse does hold true if the sequence is monotone.

Theorem 3.3.7 A bounded monotone sequence converges.

Proof. We present the proof for bounded monotone increasing sequences. The proof for bounded monotone decreasing sequences is similar and is left as an exercise (see Exercise 3.14).

Let $\{a_n\}$ be monotone increasing. To show that it converges, we need a candidate for the limit value L of the sequence. There is really only one – the least upper bound L for the sequence $\{a_n\}$. We will show that as the terms a_n increase toward L, they get arbitrarily close to this value for large n.

So, let L be the least upper bound for $\{a_n\}$. According to Theorem 3.3.5, it satisfies two proprieties:

1. $a_n \leq L$ for every n (that is, L is an upper bound), and
2. for any given $\varepsilon > 0$, there is an integer $N = N(\varepsilon)$ for which $a_N > L - \varepsilon$.

Now, since $\{a_n\}$ is monotone increasing, we have that for all $n > N$, $a_n \geq a_N$. This shows that for all $n > N$,

$$L - \varepsilon < a_N \leq a_n \leq L < L + \varepsilon; \text{ that is, } |a_n - L| < \varepsilon.$$

This proves that $\{a_n\}$ converges to its least upper bound L.   ∎

Theorem 3.3.8 A monotone increasing sequence either converges or diverges to ∞. A monotone decreasing sequence either converges or diverges to $-\infty$.

Proof. Let $\{a_n\}$ be monotone increasing. If it is bounded, it is convergent, by Theorem 3.3.7. If it is not bounded, no matter which α we pick, α cannot be an upper bound. This means that for any α, we will be able to find an N for which $a_N > \alpha$. Since $\{a_n\}$ is monotone increasing, this means that for all $n > N$, $a_n > \alpha$. By definition 3.2.3, this means $\{a_n\}$ diverges to ∞.

The proof for monotone decreasing sequences is left as an exercise (see Exercise 3.15).   ∎

■ **Example 3.14** If $r > 1$, the sequence $\{a_n\} = \{r^n\}$ diverges to ∞.

We have, $a_{n+1} = r^{n+1} = r a_n$. Since $r > 1$, this means that, for all n, $a_{n+1} \geq a_n$, showing that the sequence $\{a_n\}$ is monotone increasing. The sequence is unbounded as, no matter which α we pick, α cannot be an upper bound (we can always find n so large that $r^n > \alpha$). Therefore, by Theorem 3.3.8, the sequence $\{r^n\}$ diverges to ∞.   ■

This result is a special case of the following theorem for the sequence $\{a_n\} = \{r^n\}$ for various values of r.

Theorem 3.3.9 Consider the sequence $\{a_n\} = \{r^n\}$, where $r \in \mathbb{R}$.
1. The sequence converges for $-1 < r \leq 1$.
2. The sequence diverges for $r > 1$ and $r \leq -1$.

Proof. In Example 3.5, we proved that the sequence converges when $|r| < 1$. The case $r = 1$ gives the constant sequence $\{1, 1, 1, \ldots\}$, which is convergent.

In Example 3.14, we proved that when $r > 1$, the sequence diverges to ∞.

When $r = -1$, the sequence is $\{-1, 1, -1, 1, \ldots\}$. In Example 3.6, we showed that it diverges.

Finally, let's consider the sequence for $r < -1$. We will use a proof by contraction to show that it diverges. Assume to the contrary that $\{r^n\}$ converges to a number L. This means that $\{r^n\}$ is bounded (see Theorem 3.3.2), and we can find numbers α and β, such that

$$\beta < r^n < \alpha, \text{ for all } n. \tag{3.3}$$

But, for $r < -1$ and n sufficiently large, we will have terms of the sequence larger than α (and other, smaller than β), which contradicts the statement in line (3.3) earlier. Thus, $\{r^n\}$ diverges.   ∎

We conclude this section with an observation. The proof of Theorem 3.3.7 shows that a bounded monotone increasing (decreasing) sequence converges to its lowest upper (greatest lower) bound. However, finding that bound is not always obvious. We give an example to illustrate that.

■ **Example 3.15** Let $\{a_n\}$ be the sequence of numbers defined by

$$a_1 = \sin(1),$$
$$a_2 = \text{the larger of } \{\sin(1), \sin(2)\},$$

$$\vdots$$

$$a_n = \text{the largest of } \{\sin(1), \sin(2), \ldots, \sin(n)\},$$

$$\vdots$$

where the angle is given in radian measure. Clearly $a_n \leq a_{n+1}$ for all n, so $\{a_n\}$ is monotone increasing. We also have $|a_n| \leq 1$ for all n, so the sequence is bounded. Thus, the sequence converges by Theorem 3.3.7, but we have not found the limit. This provides an example where we know the limit of the sequence exists but we don't know its value. It may seem that the limit is 1, but that isn't easy to prove, as we don't know if 1 is the least upper bound. ■

Exercise 3.9 Prove part 1 of Theorem 3.3.4 using a proof by contradiction. That is, start your proof by assuming that $\lim a_n \neq 0$.

Exercise 3.10 Prove that if α is a lower bound for the sequence of real numbers $\{x_n\}$, then α is a lower bound for $\{|x_n|\}$.

Exercise 3.11 Prove that if β is an upper bound for the sequence of real numbers $\{|x_n|\}$, then β is an upper bound for $\{x_n\}$. ∎

Exercise 3.12 Provide a detailed proof for Theorem 3.3.6. Follow the proof of Theorem 3.3.5 and make appropriate adjustments. ∎

Exercise 3.13 Prove the result regarding lower bounds for Theorem 3.3.3. That is, prove that if a sequence $\{a_n\}$ converges to L and β is a lower bound for $\{a_n\}$ (that is, $\beta \leq a_n$ for all n), then $\beta \leq L$. ∎

Exercise 3.14 Prove that a bounded monotone decreasing sequence converges. *Hint.* Make appropriate modifications to the proof of Theorem 3.3.7 to prove that the sequence converges to its greatest lower bound. ∎

Exercise 3.15 Prove that a monotone decreasing sequence either converges or diverges to $-\infty$. ∎

Exercise 3.16 Show that each of the sequences below is bounded and monotone. You already saw some of them in the examples. Conclude, using Theorem 3.3.7, that the sequences converge.
1. $a_n = \frac{n^2}{2^n}$;
2. $a_n = 0.123\ldots n$;
3. $a_n = \sqrt{n+1} - \sqrt{n}$;
4. $a_1 = \sqrt{2}, a_{n+1} = \sqrt{2 + a_n}$;
5. $a_0 = 0, a_1 = 1$. For $n \geq 2, a_n = \frac{1 + a_{n-1}}{2 + a_{n-2}}$. ∎

Exercise 3.17 Let $\{a_n\}$ be a sequence with $\lim a_n = a$. Construct a new sequence $\{b_n\}$ by removing or changing the values of a *finite* number of terms of $\{a_n\}$. Show that the sequence $\{b_n\}$ converges and $\lim b_n = a$. ∎

4 Combinations of Sequences

We now concentrate on the proofs of theorems about combinations of sequences.

> **Theorem 3.4.1** Let $\{a_n\}$ and $\{b_n\}$ be convergent sequences of real numbers with $\lim a_n = a$ and $\lim b_n = b$. Then $\{a_n + b_n\}$ converges to $a + b$.

Proof. This is one of the most simple theorems to prove regarding the behavior of combinations of sequences, but it usually takes some time for the proof to sink in. We give the major ideas of how to prove a theorem of this type.

First, it is often helpful to state what you know and what you want to show. Here, we know that both $|a_n - a|$ and $|b_n - b|$ can be made small for sufficiently large n. We want to show that $|(a_n + b_n) - (a + b)|$ can be made small for sufficiently large n. So, the challenge is to write $|(a_n + b_n) - (a + b)|$ in terms of $|a_n - a|$ and $|b_n - b|$.

We can write

$$|(a_n + b_n) - (a + b)| = |(a_n - a) + (b_n - b)|,$$

and apply the triangle inequality (see Section 6.3 of the Appendix), which implies

$$|(a_n - a) + (b_n - b)| \le |(a_n - a)| + |(b_n - b)|.$$

Let's take $\varepsilon > 0$. We know we can make $|(a_n - a)|$ and $|(b_n - b)|$ as small as we like, by making n sufficiently large. If we make each be less than $\frac{\varepsilon}{2}$, we will have what we need. But we need to be a little bit careful. The different sequences may require us to go to different terms to get what we need. So, we know there is a number N_1, so that if $n > N_1$, then $|(a_n - a)| < \frac{\varepsilon}{2}$. There is also a number N_2, so that if $n > N_2$, then $|(b_n - b)| < \frac{\varepsilon}{2}$. We want both to happen, and if we take $N = \max\{N_1, N_2\}$, both do happen for $n > N$. So, for any $\varepsilon > 0$, we have now found N, such that

$$|(a_n + b_n) - (a + b)| = |(a_n - a) + (b_n - b)| \le |(a_n - a)| + |(b_n - b)|$$
$$< \frac{\varepsilon}{2} + \frac{\varepsilon}{2}$$
$$= \varepsilon, \text{ for all } n > N.$$

This proves that $\{a_n + b_n\}$ converges to $a + b$. ∎

After you feel that you understand this, you should try to rework it without referring to notes. Focus on the main ideas. Remember, understanding the definition of convergence and how to use it is key!

A modification of Theorem 3.4.1 can be used to prove that if $\lim a_n = a$ and $\lim b_n = b$, then $\lim(a_n - b_n) = a - b$ (see Exercise 3.21).

The proof of the next theorem is a bit more sophisticated in terms of the algebraic manipulations it requires, but we use the same idea as in the proof of Theorem 3.4.1.

> **Theorem 3.4.2** If $\{a_n\}$ and $\{b_n\}$ are convergent sequences of real numbers with $\lim a_n = a$ and $\lim b_n = b$. Then $\lim a_n b_n = ab$.

Proof. We begin the same way. We know that both $|a_n - a|$ and $|b_n - b|$ can be made small for large n, and want to prove that $|a_n b_n - ab|$ can be made small. Here the algebra is a bit trickier. By the triangle inequality, we have

$$
\begin{aligned}
|a_n b_n - ab| &= |a_n b_n - ab_n + ab_n - a_n b_n| \\
&\leq |a_n b_n - ab_n| + |ab_n - a_n b_n| \\
&\leq |b_n||a_n - a| + |a||b_n - b|.
\end{aligned}
$$

So, we have done some good things. One of the terms on the last line has $|a_n - a|$ and the other one has $|b_n - b|$. If pick an $\varepsilon > 0$, we want to make

$$
|b_n||a_n - a| < \frac{\varepsilon}{2} \text{ and } |a||b_n - b| < \frac{\varepsilon}{2} \text{ for sufficiently large } n.
$$

The term $|a||b_n - b|$ is no problem. Just take n large enough to make

$$
|b_n - b| < \frac{\varepsilon}{2|a|}, \tag{3.4}
$$

when $a \neq 0$. More formally, there is a number N_1 so that, if $n > N_1$, the condition in Equation (3.4) is satisfied. Note that if $a = 0$, then we already have $|a||b_n - b| = 0 < \frac{\varepsilon}{2}$.

The term $|b_n||a_n - a|$ requires more care. First, since $\{b_n\}$ is a convergent sequence it is bounded (Theorem 3.3.2). Thus, there is a number K so that $|b_n| < K$ for all n.

Next, since $\{a_n\}$ converges to a, we can make

$$
|a_n - a| < \frac{\varepsilon}{2K}, \tag{3.5}
$$

by making n sufficiently large. More formally, there is a number N_2 so that if $n > N_2$, Equation (3.5) is satisfied.

Now take $N = \max\{N_1, N_2\}$. Then both inequalities hold for all $n > N$, and we have

$$
|a_n b_n - ab| \leq |a_n b_n - ab_n| + |ab_n - a_n b_n| < \frac{\varepsilon}{2} + \frac{\varepsilon}{2} < \varepsilon.
$$

This proves that $\{a_n b_n\}$ converges to ab. ■

The proof of our next theorem requires the following preliminary result.

Proposition 3.4.3 If $\{b_n\}$ is a sequence of nonzero numbers that converges to b, and $b \neq 0$, the sequence $\{\left|\frac{1}{b_n}\right|\}$ is bounded.

Proof: Since b_n are nonzero, we have $\left|\frac{1}{b_n}\right| > 0$. Thus, we need to show that $\left|\frac{1}{b_n}\right|$ cannot get infinitely large.

Let $\{b_n\}$ converge to $b \neq 0$. This means that given any $\varepsilon > 0$, we can find $N = N(\varepsilon)$, such that for all $n > N(\varepsilon)$ we have $b - \varepsilon < b_n < b + \varepsilon$. Because $b \neq 0$, we can take $\varepsilon > 0$ so small that the interval $(b - \varepsilon, b + \varepsilon)$ does not contain 0. That is, either both $b - \varepsilon$ and $b + \varepsilon$ are positive, or they are both negative. If they are both positive, we have $0 < b - \varepsilon < |b_n| < b + \varepsilon$. If they are both negative, we get $0 < |b + \varepsilon| \leq |b_n| \leq |b - \varepsilon|$. Now, if we take $M = \min\{|b - \varepsilon|, |b + \varepsilon|\}$, we have $0 < M < |b_n|$. Therefore

$$\left|\frac{1}{b_n}\right| < \frac{1}{M}, \quad \text{for all } n > N(\varepsilon).$$

But the set of all $\left|\frac{1}{b_n}\right|$ with $n \leq N(\varepsilon)$ is finite, and thus bounded (see Theorem 3.3.1). If K is an upper bound for this set, then for $\alpha = \max\{\frac{1}{M}, K\}$ we will have

$$\left|\frac{1}{b_n}\right| \leq \alpha, \quad \text{for all } n.$$

This proves that $\left|\frac{1}{b_n}\right|$ is bounded.

Theorem 3.4.4 If $\{b_n\}$ is a sequence of nonzero numbers that converges to b, and $b \neq 0$, then $\{\frac{1}{b_n}\}$ converges to $\frac{1}{b}$.

Proof. Knowing that we can make $|b_n - b|$ as small as we want for large n, we want to show that $\left|\frac{1}{b_n} - \frac{1}{b}\right|$ can also be made as small as we want for large enough n.

We have

$$\left|\frac{1}{b_n} - \frac{1}{b}\right| = \frac{|b_n - b|}{|bb_n|}.$$

We know, from Proposition 3.4.3 that $\{\left|\frac{1}{b_n}\right|\}$ is bounded. Since $b \neq 0$, so is the sequence $\{\left|\frac{1}{bb_n}\right|\}$. Therefore, there is a number K so that

$$\left|\frac{1}{bb_n}\right| < K, \quad \text{for all } n.$$

Now, let $\varepsilon > 0$. Since $\lim b_n = b$, there is an $N = N(\varepsilon)$, so that if $n > N$, then

$$|b_n - b| < \frac{\varepsilon}{K}.$$

This makes

$$\left| \frac{1}{b_n} - \frac{1}{b} \right| = \frac{|b_n - b|}{|bb_n|} < K|b_n - b| < K\frac{\varepsilon}{K} = \varepsilon \text{ for } n > N.$$

Thus, we have proved that $\{\frac{1}{b_n}\}$ converges to $\frac{1}{b}$. ∎

Exercise 3.18 Prove that if the sequence $\{a_n\}$ converges to a and c is a constant, then $\{ca_n\}$ converges to ca. ∎

Exercise 3.19 Prove that $\{\frac{3n+1}{2n}\}$ converges to $\frac{3}{2}$. *Hint.* Divide both the numerator and denominator by n, and use that $\lim \frac{1}{n} = 0$. ∎

Exercise 3.20 Prove that $\{-\frac{5n^2}{7n^2+n-1}\}$ converges to $-\frac{5}{7}$. *Hint.* Divide both the numerator and denominator by n^2, and use that $\lim \frac{1}{n^2} = 0$. ∎

Exercise 3.21 Let $\{a_n\}$ and $\{b_n\}$ be convergent sequences of real numbers with $\lim a_n = a$ and $\lim b_n = b$. Then $\{a_n - b_n\}$ converges to $a - b$. ∎

Exercise 3.22 Give an example of two divergent sequences whose sum converges. ∎

Exercise 3.23 Give examples of sequences $\{a_n\}$ and $\{b_n\}$ that both diverge to ∞, for which:
1. $\{a_n - b_n\}$ diverges to ∞;
2. $\{a_n - b_n\}$ converges to 5. ∎

Infinite Series of Real Numbers

In this section we consider series of numbers, a topic that is intimately related to sequences. What one would like to do in a series is to add an infinite number of terms. Such "infinite sums" are called *infinite series* (or just series). We begin with a sequence $\{a_n\}$ and want to consider the sum of its (infinitely many) terms. That is, we would like to look at

$$a_1 + a_2 + a_3 + \cdots = \sum_{n=1}^{\infty} a_n.$$

How do we decide what the value of this sum is, or if it even exists? To answer this, construct a new sequence $\{S_n\}$ as follows:

$$S_1 = a_1,$$
$$S_2 = a_1 + a_2,$$
$$S_3 = a_1 + a_2 + a_3,$$
$$\cdots$$
$$S_n = a_1 + a_2 + \cdots + a_n.$$

Definition 3.5.1 For an infinite series

$$\sum_{n=1}^{\infty} a_n, \tag{3.6}$$

the sequence $\{S_n\}$, where $S_n = a_1 + a_2 + \cdots + a_n$, is called *the sequence of partial sums* for that infinite series.

The next definition tells us how to interpret infinite sums of this type.

Definition 3.5.2 If the sequence of partial sums $\{S_n\}$ from Definition 3.5.1 converges to a number L, we write

$$L = \sum_{n=1}^{\infty} a_n,$$

and we say that the infinite series *converges* to L. When such a number does not exist, or when $\lim S_n = \pm\infty$, we say that the infinite series $\sum_{n=1}^{\infty} a_n$ *diverges*.

Definition 3.5.2 tells us that an infinite series converges when its sequence of partial sums is a convergent sequence. The following observation is based on that connection.

Recall that if you remove a finite number from the terms of a convergent sequence, the new sequence converges to the same limit (see Exercise 3.17). In the case of infinite series, once can easily show (see Exercise 3.24) that removing any finite number of terms in an infinite series will not affect its convergence/divergence. Of course, in general, this will affect *the value* of the limit for the convergent series. Similarly, adding any finite number of terms anywhere in an infinite series will not affect its convergence/divergence.

In particular, if we consider an infinite series $\sum_{n=1}^{\infty} a_n$ and remove the first N terms, the new infinite series

$$a_N + a_{n+1} + a_{N+2} + \cdots = \sum_{n=N}^{\infty} a_n, \tag{3.7}$$

will converge when $\sum_{n=1}^{\infty} a_n$ converges and diverge when $\sum_{n=1}^{\infty} a_n$ diverges. We will refer to the series $\sum_{n=N}^{\infty} a_n$ as the *tail* of the series $\sum_{n=1}^{\infty} a_n$.

A focus of this section is how to determine if an infinite series converges or diverges. The following result is a convenient way to show that a series diverges.

> **Theorem 3.5.1** (*Divergence Test*) If $\lim a_n \neq 0$, then $\sum_{n=1}^{\infty} a_n$ diverges.

Proof. We will prove the equivalent contrapositive statement (see Theorem 1.2.2). That is, we will prove that if $\sum_{n=1}^{\infty} a_n$ converges, then $\lim a_n = 0$.

Let $\sum_{n=1}^{\infty} a_n$ converge to L. Since $S_n = a_1 + a_2 + \cdots + a_n$ and $S_{n-1} = a_1 + a_2 + \cdots + a_{n-1}$, we have $a_n = S_n - S_{n-1}$. Thus,

$$\lim a_n = \lim (S_n - S_{n-1}) = \lim S_n - \lim S_{n-1} = L - L = 0.$$

∎

■ **Example 3.16** Consider the series

$$\sum_{n=1}^{\infty} a_n = \sum_{n=1}^{\infty} \frac{3n}{2n+1}.$$

Since $\lim a_n = \lim \frac{3n}{2n+1} = \frac{3}{2} \neq 0$, the series diverges. ■

Note that Theorem 3.5.1 is only an *if-then* statement and NOT an *if-and-only-if* statement. In fact, the converse statement is false – there are many series for which $\lim a_n = 0$ but the series $\sum_{n=1}^{\infty} a_n$ diverges.

We now define a special series that plays a very important role in the theory of infinite series.

Definition 3.5.3 Let a and r be real numbers. A series of the form

$$a + ar + ar^2 + ar^3 + \cdots = \sum_{n=0}^{\infty} ar^n \tag{3.8}$$

is called a *geometric series*.

Geometric series are important for two reasons: (1) They are a type of series for which, when convergent, we can easily find their sum and (2) They are used to derive more general criteria for convergence/divergence of infinite series.

Theorem 3.5.2 Consider the geometric series $\sum_{n=0}^{\infty} ar^n$.

1. If $a = 0$, the series converges to zero.

2. If $a \neq 0$, the series diverges for $|r| \geq 1$ and converges to $\frac{a}{1-r}$ for $|r| < 1$.

Proof. 1. When $a = 0$, $S_n = 0$ for all n, so $\lim S_n = 0$.

2. Now let $a \neq 0$. We will look at the sequence $\{S_n\}$ of partial sums to determine when that sequence converges. The n-th partial sum is

$$S_n = a + ar + ar^2 + \cdots + ar^{n-1},$$

so

$$rS_n = ar + ar^2 + \cdots + ar^{n-1} + ar^n.$$

Then

$$S_n - rS_n = (1 - r)S_n = a - ar^n.$$

When $r \neq 1$,

$$S_n = a \frac{1 - r^n}{1 - r}. \tag{3.9}$$

Recall now Theorem 3.3.9. If $|r| < 1$, $\lim r^n = 0$ and, from Equation (3.9),

$$\lim S_n = \frac{a}{1 - r}.$$

For $|r| > 1$ and $r = -1$, we know the sequence $\{r^n\}$ diverges (see again Theorem 3.3.9), so the geometric series diverges. When $r = 1$, $S_n = \underbrace{a + a + \cdots + a}_{n} = na$. Thus, $\{S_n\}$ diverges in this case as well, and so does the geometric series. ∎

Our next example illustrates how the geometric series may help to solve practical problems. We include this just as an illustration, since our focus here is on developing theoretical proofs.

■ **Example 3.17** A ball is dropped from a height of 10 feet. Each time it bounces it rebounds to $\frac{3}{4}$ the height of the previous bounce. How far does the ball travel before it comes to rest?

We calculate separately the distance traveled by the ball when it is moving up and the distance when it is moving down.

The distance of "down" travel is

$$10 + \left(\frac{3}{4}\right)10 + \left(\frac{3}{4}\right)^2 10 + \left(\frac{3}{4}\right)^3 10 + \cdots = \frac{10}{1 - \frac{3}{4}} = \frac{10}{\frac{1}{4}} = 40.$$

The distance of "up" travel is

$$7.5 + \left(\frac{3}{4}\right)7.5 + \left(\frac{3}{4}\right)^2 7.5 + \left(\frac{3}{4}\right)^3 7.5 + \cdots = \frac{7.5}{1 - \frac{3}{4}} = \frac{7.5}{\frac{1}{4}} = 30.$$

Thus, the ball travels $40 + 30 = 70$ feet. ■

Since, by definition, an infinite series converges when the sequence of partial sums converges, the property listed below follows directly from properties we proved in Section 3.4. The proof is left as exercises (see Exercise 3.25).

> **Theorem 3.5.3** Suppose the series $\sum_{n=1}^{\infty} a_n$ converges to a and $\sum_{n=1}^{\infty} b_n$ converges to b. Then, for any numbers c and d, $\sum_{n=1}^{\infty} (ca_n + db_n)$ converges to $ca + db$.

There are several other tests for convergence/divergence of series, some of which we present next.

> **Theorem 3.5.4** (Comparison Test) Let $\{a_n\}$ and $\{b_n\}$ be sequences of positive numbers that satisfy the condition
>
> $a_n \leq b_n$, for every n.
>
> 1. If $\sum_{n=1}^{\infty} b_n$ converges, then $\sum_{n=1}^{\infty} a_n$ converges.
> 2. If $\sum_{n=1}^{\infty} a_n$ diverges, then $\sum_{n=1}^{\infty} b_n$ diverges.

Proof. Let A_n and B_n be the partial sums of $\sum_{n=1}^{\infty} a_n$ and $\sum_{n=1}^{\infty} b_n$. Since for all n, $a_n > 0$, $b_n > 0$, and $a_n \leq b_n$, the sequences $\{A_n\}$ and $\{B_n\}$ are monotone increasing with $A_n \leq B_n$.

Proof of Part 1. Let $\sum_{n=1}^{\infty} b_n$ converge to B. This means that $\lim B_n = B$. Since $\{B_n\}$ is monotone increasing, $B_n \leq B$ for all n. We thus have $0 < A_n \leq B_n \leq B$, which shows that the sequence $\{A_n\}$ is bounded. As we know that a bounded monotone increasing sequence converges (see Theorem 3.3.7), this shows that $\sum_{n=1}^{\infty} a_n$ converges.

Proof of Part 2. Let $\sum_{n=1}^{\infty} a_n$ diverge. Equivalently, the monotonic increasing sequence $\{A_n\}$ diverges, which means (see Theorem 3.3.8) it diverges to ∞. And since $A_n \leq B_n$ for all n, the monotone increasing sequence B_n also diverges to ∞. Therefore $\sum_{n=1}^{\infty} b_n$ diverges. ∎

The next result is very important and will be used in the proofs of the theorems that follow.

Theorem 3.5.5 Consider an infinite series $\sum_1^{\infty} a_n$.

1. If $\sum_1^{\infty} |a_n|$ converges, then $\sum_1^{\infty} a_n$ also converges.

2. Equivalently, if $\sum_1^{\infty} a_n$ diverges, then $\sum_1^{\infty} |a_n|$ diverges.

Proof. The equivalence of the two statements follows immediately since statement 2 is the contrapositive of statement 1 (see Theorem 1.2.2). We will prove Statement 1.

Let $\sum_1^{\infty} |a_n|$ converge. Note that, for all n, we have either $|a_n| = a_n$ or $|a_n| = -a_n$, which gives

$$0 \leq a_n + |a_n| \leq 2|a_n|.$$

Now, since $\sum_1^{\infty} |a_n|$ converges, the Comparison Test shows that $\sum_1^{\infty}(a_n + |a_n|)$ also converges. We then have

$$\sum_1^{\infty} a_n = \sum_1^{\infty}(a_n + |a_n|) - \sum_1^{\infty} |a_n|.$$

Thus, $\sum_1^{\infty} a_n$ is a difference of two convergent series. Theorem 3.5.3 now implies that $\sum_1^{\infty} a_n$ converges. ∎

Definition 3.5.4 Consider the series $\sum_1^{\infty} a_n$. If the series of absolute values $\sum_1^{\infty} |a_n|$ converges, We say that $\sum_1^{\infty} a_n$ is *absolutely convergent*.

With this definition, Theorem 3.5.5 can be stated in the following way:

Every absolutely convergent series converges.

The following tests are widely used to determine if a series converges or diverges.

Theorem 3.5.6 (*The Ratio Test*) Consider the infinite series $\sum_{n=1}^{\infty} a_n$, and let

$$\lim \frac{|a_{n+1}|}{|a_n|} = r.$$

1. If $r < 1$, the series converges;
2. If $r > 1$, the series diverges;

For $r = 1$, the test in inconclusive.

Proof. We know that $\lim \frac{|a_{n+1}|}{|a_n|} = r$. In the case $r > 0$, this means that for any sufficiently small $\varepsilon > 0$, we will have $r > \varepsilon > 0$, and we can find $N = N(\varepsilon)$, such that

$$0 < r - \varepsilon < \frac{|a_{n+1}|}{|a_n|} < r + \varepsilon, \text{ for all } n \geq N. \tag{3.10}$$

Proof of 1. Let $0 < r < 1$. We can now take $\varepsilon > 0$ so small that $q = r + \varepsilon < 1$. Thus, for all $n \geq N$, we will have

$$0 < |a_{N+1}| < |a_N|q.$$

Note that this means that $|a_{N+2}| < |a_{N+1}|q < |a_N|q^2$; $|a_{N+3}| < |a_{N+2}|q < |a_N|q^3$, and so on. Therefore, for all $n \geq N$, we have that

$$0 < |a_{N+k}| < |a_N|q^k, \ k = 1, 2, 3, \dots$$

Since $q < 1$, the geometric series $\sum_{k=1}^{\infty} |a_n|q^k$ converges. Thus, by the comparison test, $\sum_{k=N}^{\infty} |a_{n+1}|$ also converges. So, we have proved that the tail of $\sum_{n=1}^{\infty} a_n$ converges (since it converges absolutely), and we know (see Exercise 3.24) that this means $\sum_{n=1}^{\infty} a_n$ converges as well.

If $r = 0$, then for any $1 > \varepsilon > 0$, we will have $\frac{|a_{n+1}|}{|a_n|} < \varepsilon$ for sufficiently large n. This means, we can find $N = N(\varepsilon)$ such that $|a_{n+1}| < |a_n|\varepsilon$, for all $n > N$. So, we have the same argument as before, with $q = \varepsilon$, and the series converges.

Proof of 2. Let $r > 1$. Then, for sufficiently large n, $\frac{|a_{n+1}|}{|a_n|} > 1$. Equivalently, we will have $|a_{n+1}| > |a_n| > 0$, for all sufficiently large n, which means the terms of the sequence $\{a_n\}$ are moving away from 0. Thus $\lim a_n \neq 0$, and $\sum_{n=1}^{\infty} a_n$ diverges by the divergence test (Theorem 3.5.1). ∎

Theorem 3.5.7 (*The Root Test*) Consider the infinite series $\sum_{n=1}^{\infty} a_n$, and let

$$\lim \sqrt[n]{|a_n|} = r.$$

1. If $r < 1$, the series converges;

2. If $r > 1$, the series diverges;

For $r = 1$, the test in inconclusive.

Proof. The proof is left as an exercise (see Exercise 3.26). ∎

Exercise 3.24 Prove that removing a *finite* number of terms from an infinite series does not affect its convergence or divergence. That is, the new series will still converge (diverge) if the original series converges (diverges).

Exercise 3.25 Prove Theorem 3.5.3. *Hint.* Consider the partial sums of each series and apply Theorem 3.4.1 and Exercise 3.18.

Exercise 3.26 Prove the Root Test using ideas similar to those we used in the poof of the Ratio Test.

Exercise 3.27 Decide whether each series converges or diverges and give the test you used.

1. $\sum \frac{1}{n^2+5n+6}$;

2. $\sum \left(\frac{n^3}{1-4n^3}\right)^n$;

3. $\sum \frac{n^4}{3n^2+6n-5}$;

4. $\sum \frac{n^n}{n!}$;

5. $\sum \frac{e^n}{n^n}$;

6. $\sum \frac{(-1)^n n!}{n^4}$.

Exercise 3.28 Show that the series $\sum_1^\infty \frac{1}{n(n+1)}$ converges to 1. *Hint.* Show that

$$S_n = \frac{1}{1\cdot 2} + \frac{1}{2\cdot 3} + \cdots + \frac{1}{n\cdot(n+1)} =$$
$$= \left(1 - \frac{1}{2}\right) + \left(\frac{1}{2} - \frac{1}{3}\right) + \left(\frac{1}{3} - \frac{1}{4}\right) + \cdots + \left(\frac{1}{n} - \frac{1}{n+1}\right).$$

6 Limit of a Function

Finding limits of functions, as the independent variable approaches a certain value or grows/decreases without a bound, is another topic you have seen in your Calculus courses. Just as in the case of limits of sequences, you have likely learned what a limit of a function means intuitively, without using a rigorous definition. In this section, we introduce the rigorous definition and expose you to how mathematical proofs for this topic can be constructed. You will go much further in a standard Real Analysis course.

Definition 3.6.1 Let $f : \mathbb{R} \to \mathbb{R}$ be a function, and I be an open interval contained in the domain of f. Let $x_0 \in I$. We say that *the limit of $f(x)$, as x approaches x_0, is the number L,* if the following is satisfied: Given any $\varepsilon > 0$, there is a $\delta = \delta(\varepsilon)$ with the property that if $|x - x_0| < \delta$, then $|f(x) - L| < \varepsilon$. In this case, we write

$$\lim_{x \to x_0} f(x) = L.$$

You will notice some similarities with Definition 3.2.2, where we defined a convergent sequence. Just as in Definition 3.2.2, we begin with picking a very small number $\varepsilon > 0$. *After ε is chosen,* we need to find a number $\delta(\varepsilon)$ that depends on ε (thus, the notation $\delta(\varepsilon)$). That number should be such that if the distance $|x - x_0| < \delta(\varepsilon)$, then the distance $|f(x) - L|$ between $f(x)$ and L is less than ε. See Figure 3.6.

■ **Example 3.18** We will show that

$$\lim_{x \to 2} x^2 = 4. \tag{3.11}$$

We choose an $\varepsilon > 0$. We need to show that there is a $\delta = \delta(\varepsilon) > 0$, such that if $|x - 2| < \delta$, then $|x^2 - 4| < \varepsilon$.

Since $x^2 - 4 = (x-2)(x+2)$, we have $|x^2 - 4| = |(x-2)||(x+2)|$. If we could say that $|x+2|$ is smaller than some constant, we would be almost done.

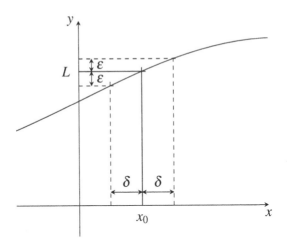

Figure 3.6: For a function $f(x)$ to be continuous at x_0, given any $\varepsilon > 0$, we should be able to find a $\delta = \delta(\varepsilon) > 0$, so that the graph of $f(x)$ lies between the lines $y = L - \varepsilon$ and $y = L + \varepsilon$ when x is in the interval $(x_0 - \delta,\ x_0 + \delta)$.

Since we know that x is close to 2, we can look at x's for which, say, $1 \leq x \leq 3$; that is x's for which $|x - 2| \leq 1$. Then $3 \leq x + 2 \leq 5$, so $|x + 2| \leq 5$. Now

$$|x^2 - 4| = |(x - 2)||(x + 2)| \leq 5|x - 2|.$$

This tells us that if we choose $\delta(\varepsilon) = \frac{\varepsilon}{5}$, and $|x - 2| < \delta(\varepsilon) = \frac{\varepsilon}{5}$, then $|x^2 - 4| < \varepsilon$. By Definition 3.6.1, we have proved that $\lim_{x \to 2} x^2 = 4$. ∎

Here is another example of this type, where we use the same strategy.

■ **Example 3.19** Show that

$$\lim_{x \to 3} x^2 - 4x + 15 = 12.$$

Let $\varepsilon > 0$ be given. We need to show that there is a $\delta = \delta(\varepsilon)$ with the following property:

If $|x - 3| < \delta$, then $|(x^2 - 4x + 15) - 12| < \varepsilon$.

We have

$$|(x^2 - 4x + 15) - 12| = |x^2 - 4x + 3| = |x - 3||x - 1|.$$

We know we can make $|x - 3|$ as small as we like. Since x is a value close to 3, we can consider a range of values for x, e.g., $2 < x < 4$, for which $|x - 1| < 3$.

This means that if we choose $\delta = \delta(\varepsilon) = \frac{\varepsilon}{3}$, then $|x - 3| < \delta$ gives us what we want:

$$|(x^2 - 4x + 15) - 12| = |x - 3||x - 1| < \left(\frac{\varepsilon}{3}\right)3 = \varepsilon,$$

which proves that $\lim_{x \to 3} x^2 - 4x + 15 = 12$. ∎

Next, we discuss limits where the function grows (decreases) to $+\infty$ ($-\infty$), as x approaches x_0.

Definition 3.6.2 Let $f : \mathbb{R} \to \mathbb{R}$ be a function, and let I be an open interval contained in the domain of f. Let $x_0 \in I$. We say that the limit of a function $f(x)$, as x approaches x_0, is infinity, if the following is satisfied: Given any number M, there is a number $\delta(M) > 0$, such that if $|x - x_0| < \delta(M)$, then $f(x) > M$. We write

$$\lim_{x \to x_0} f(x) = \infty,$$

and we say that the line $x = x_0$ is a *vertical asymptote*.

■ **Example 3.20** Show that

$$\lim_{x \to 5} \frac{1}{(x - 5)^2} = \infty.$$

Note that the function is not defined for $x = 5$ but is defined for any x arbitrarily close to 5. Intuitively, when x gets close to 5, the denominator is getting very small, thus $\frac{1}{(x-5)^2}$ is getting very large. Now, let's *prove* that the limit is ∞ by using Definition 3.6.2.

Choose a large number M. We want to find $\delta(M)$, which will guarantee that $\frac{1}{(x-5)^2} > M$ whenever $|x - 5| < \delta(M)$. But $\frac{1}{(x-5)^2} > M$ is the same as $(x - 5)^2 < \frac{1}{M}$, and the same as $|x - 5| < \frac{1}{\sqrt{M}}$. This tells us that we may choose $\delta(M) = \frac{1}{\sqrt{M}}$. Now for all x that satisfy $|x - 5| < \frac{1}{\sqrt{M}}$,

$$\frac{1}{(x - 5)^2} > \frac{1}{\left(\frac{1}{\sqrt{M}}\right)^2} = M.$$

By Definition 3.6.2, we have proved that $\lim_{x \to 5} \frac{1}{(x-5)^2} = \infty$ and that the line $x = 5$ is a vertical asymptote. ∎

If we consider values of x that get close to x_0 only by values $x > x_0$, Definitions 3.6.1 and 3.6.2 describe one-sided *limits from the right*. We then write

$$\lim_{x \to x_0^+} f(x) = L \quad \text{and} \quad \lim_{x \to x_0^+} f(x) = \infty.$$

If x approaches x_0 only by values $x < x_0$, we obtain *limits from the left*:

$$\lim_{x \to x_0^-} f(x) = L \quad \text{and} \quad \lim_{x \to x_0^-} f(x) = \infty.$$

■ **Example 3.21** let $f(x) = \frac{1}{2-x}$. This function is not defined for $x = 2$ but is defined for any x arbitrarily close to 2. Intuitively, as x approaches 2 from the left, the denominator $2 - x$ approaches zero by positive values, thus $f(x)$ gets indefinitely larger in the positive direction (see Figure 3.7). It appears, the following should be true.

$$\lim_{x \to 2^-} \frac{1}{2-x} = \infty.$$

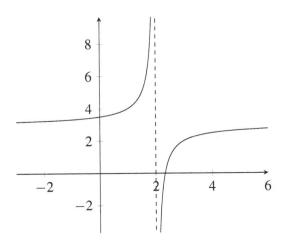

Figure 3.7: The graph of $f(x) = \frac{1}{2-x}$ near the point $x = 2$. The function is not defined for $x = 2$. It grows without a bound as $x \to 2^-$ and decreases without a bound as $x \to 2^+$. The line $x = 2$ is a vertical asymptote. Note that the two-sided limit $\lim_{x \to 2} f(x)$ does not exist in this case.

Now, let's *prove* that $f(x)$ approaches ∞, using Definition 3.6.2. We chose a very large $M > 0$. We need to show there is a positive number $\delta(M)$, so that if $x < 2$ and $|x - 2| < \delta(M)$, then $\frac{1}{2-x} > M$. But this happens if and only if $\frac{1}{M} > 2 - x$, so we may take $\delta(M) = \frac{1}{M}$.   ■

Exercise 3.29 Using Definitions 3.6.1 and 3.6.2, prove that

1. $\lim_{x \to 4}(x^2 - 16) = 0$;
2. $\lim_{x \to -3}(x^2 - 9) = 0$;
3. $\lim_{x \to 1}(x^2 - x - 2) = -2$;
4. $\lim_{x \to 4}(x^2 - 3x + 8) = 12$;
5. $\lim_{x \to 0} e^{\frac{1}{|x|}} = \infty$.

Exercise 3.30 Prove that for the function in Example 3.21 (see also Figure 3.7), we have

$$\lim_{x \to 2^+} \frac{1}{2 - x} = -\infty.$$

Exercise 3.31 Show that

$$\lim_{x \to 7} \frac{-3}{(2x - 14)^2} = -\infty.$$

Exercise 3.32 Prove that

$$\lim_{x \to -\frac{1}{3}^-} \frac{2 - x}{3x + 1} = -\infty,$$

and that

$$\lim_{x \to -\frac{1}{3}^+} \frac{2 - x}{3x + 1} = \infty.$$

Exercise 3.33 Show that

$$\lim_{x \to 4^+} \frac{x + 1}{x - 4} = \infty,$$

and that

$$\lim_{x \to 4^-} \frac{x+1}{x-4} = -\infty.$$

3.7 Limit of a Function for $x \to \infty$

Until now, we considered limits of functions when the independent variable x approaches a *finite* value x_0. The next definition addresses the case when x *approaches* ∞.

Definition 3.7.1 Let $f : \mathbb{R} \to \mathbb{R}$ be a function. We say that the limit of f, as x approaches ∞ (or, as x grows without a bound) is L, if the following is satisfied: For any $\varepsilon > 0$, there is a number $N = N(\varepsilon) > 0$, such that,

if $x > N$, then $|f(x) - L| < \varepsilon$.

The number L is called a *horizontal asymptote*. We write

$$\lim_{x \to \infty} f(x) = L.$$

To translate this, it means that, for sufficiently large x, you can make $f(x)$ arbitrarily close to L (within distance ε, no matter how small $\varepsilon > 0$ is). We give an example.

■ **Example 3.22** Using Definition 3.7.1, show that

$$\lim_{x \to \infty} \frac{3x}{1+x} = 3.$$

First, let's see what our intuition tells us. When x is very large, $x + 1 \approx x$ Thus, for very large x we will have

$$\lim_{x \to \infty} \frac{3x}{1+x} \approx \frac{3x}{x} = 3.$$

The graph of the function $f(x) = \frac{3x}{1+x}$ (see Figure 3.8), also appears to suggest that $y = 3$ is a horizontal asymptote.

Now, let's *prove* this, using the definition.

Let $\varepsilon > 0$. We want to find $N = N(\varepsilon)$ so that

$$\text{for } x > N(\varepsilon), \text{ we have } \left| \frac{3x}{1+x} - 3 \right| < \varepsilon. \tag{3.12}$$

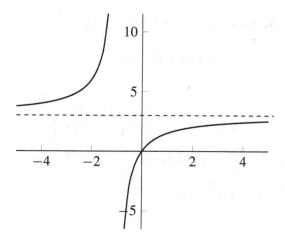

Figure 3.8: Graph of $f(x) = \frac{3x}{1+x}$. The line $y = 3$ is a horizontal asymptote.

But

$$\left| \frac{3x}{1+x} - 3 \right| = \left| \frac{3x - 3 - 3x}{1+x} \right| = \left| \frac{-3}{1+x} \right| = \frac{3}{1+x}.$$

We can drop the absolute values because, as $x \to \infty$, we are looking at large positive x-values. We want

$$\frac{3}{1+x} < \varepsilon, \text{ for sufficiently large } x > N(\varepsilon).$$

Now,

$$\frac{3}{1+x} < \varepsilon \text{ is the same as } \frac{1}{1+x} < \frac{\varepsilon}{3}, \text{ or } x > \frac{3}{\varepsilon} - 1.$$

So, if we take $N = N(\varepsilon) = \frac{3}{\varepsilon} - 1$, the desired condition from Equation (3.12) is satisfied. This proves that

$$\lim_{x \to \infty} \frac{3x}{1+x} = 3.$$

∎

■ **Example 3.23** Show that

$$\lim_{x \to \infty} \frac{-5x^2 - 2}{x^2 + 3} = -5.$$

Let $\varepsilon > 0$. We need to find $N = N(\varepsilon)$, so that

$$\left| \frac{-5x^2 - 2}{x^2 + 3} - (-5) \right| < \varepsilon, \text{ for all } x > N. \tag{3.13}$$

But

$$\left| \frac{-5x^2 - 2}{x^2 + 3} - (-5) \right| = \left| \frac{-5x^2 - 2 + 5x^2 + 15}{x^2 + 3} \right| = \left| \frac{13}{x^2 + 3} \right|.$$

So, we want to find $N = N(\varepsilon)$ such that

$$\left| \frac{13}{x^2 + 3} \right| < \varepsilon, \text{ for all } x > N.$$

Now

$$\left| \frac{13}{x^2 + 3} \right| < \varepsilon \text{ is the same as } x^2 > \frac{13 - 3\varepsilon}{\varepsilon}, \text{ which is the same as,}$$

$$x > \sqrt{\frac{13 - 3\varepsilon}{\varepsilon}}.$$

So, if we choose $N = N(\varepsilon) = \sqrt{\frac{13-3\varepsilon}{\varepsilon}}$, the condition from Equation (3.13) will be satisfied. ∎

Exercise 3.34 Following closely Definition 3.7.1, see if you can define what it would mean for

$$\lim_{x \to -\infty} f(x) = L$$

You should compare your answer to that in the textbooks [1] or [2] (see the bibliography at the end of this chapter) and check if you were correct. Be sure to congratulate yourself if you got it right and to spend some time to understand what you may have missed, if you didn't. ∎

Exercise 3.35 Looking at Definitions 3.6.2 and 3.7.1, see if you can give the rigorous definition for each of the following:
   1. $\lim_{x \to \infty} f(x) = \infty$;
   2. $\lim_{x \to -\infty} f(x) = \infty$;
   3. $\lim_{x \to \infty} f(x) = -\infty$;
   4. $\lim_{x \to -\infty} f(x) = -\infty$.

   You should compare each of your answers to the definitions given in the textbooks [1] or [2] (see the bibliography at the end of this chapter) and check if you were correct. Be sure to congratulate yourself if you got all of them right and to spend some time to understand what you may have missed, if you didn't. ∎

Exercise 3.36 Using the definition from Exercise 3.34, prove that

$$\lim_{x \to -\infty} \frac{4x}{2-x} = -4.$$

∎

Exercise 3.37 Using the definition from Exercise 3.34, prove that

$$\lim_{x \to -\infty} \frac{2+3x^2}{2x^2-1} = \frac{3}{2}.$$

∎

Exercise 3.38 Working with the definitions from Exercise 3.35, prove that:

1. $\lim_{x \to -\infty} \frac{-3x^2+7x-1}{2x+5} = \infty$;

2. $\lim_{x \to \infty} \frac{-3x^2+7x-1}{2x+5} = -\infty$;

3. $\lim_{x \to \infty} \frac{2x^3-4}{x-8} = \infty$;

4. $\lim_{x \to -\infty} \frac{2x^3-4}{x-8} = \infty$.

∎

8 Continuous Functions

In your calculus courses, you have learned how to recognize from the graph of a function whether it is continuous or not. Heuristically, a function is continuous over an interval within its domain, if it can be graphed over this interval without lifting the pen off the paper. This is not a mathematical definition, however, and we now move on to making this notion mathematically rigorous.

Definition 3.8.1 (*Continuity at a point*) Let f be a function with domain $\mathscr{D}(f)$ and $I \subseteq \mathscr{D}(f)$ be an open interval. We say that the function f is *continuous at a point* $x_0 \in I$ if, given $\varepsilon > 0$, there is a $\delta = \delta(\varepsilon) > 0$, such that

$$|f(x) - f(x_0)| < \varepsilon, \text{ for all } x \in I \text{ that satisfy } |x - x_0| < \delta.$$

Definition 3.8.2 (*Continuity over an interval*) Let f be a function with domain $\mathscr{D}(f)$ and $I \subseteq \mathscr{D}(f)$ be an open interval. We say that the function f is *continuous over the interval* I, if f is continuous at every point $x \in I$.

Compare Definition 3.8.1 with Definition 3.6.1. The only difference is that for continuity at x_0, we require that $L = \lim_{x \to x_0} f(x) = f(x_0)$; that is, we want the limit L to be the function value at x_0 (see Figure 3.9).

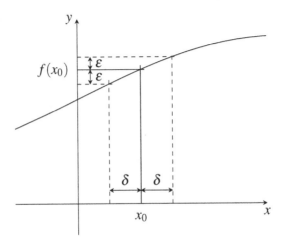

Figure 3.9: For a function $f(x)$ to be continuous at x_0, its graph has to lie between the lines $y = f(x_0) - \varepsilon$ and $y = f(x_0) + \varepsilon$ when x is in the interval $(x_0 - \delta, x_0 + \delta)$.

■ **Example 3.24** Show that the function $f(x) = x^2$ is continuous at any point $x_0 \in \mathbb{R}$.

Pick any $x_0 \in \mathbb{R}$. We have to prove that $\lim_{x \to x_0} x^2 = x_0^2$.

Let $\varepsilon > 0$. We need to find a number $\delta = \delta(\varepsilon)$, such that if $|x - x_0| < \delta$, then $|f(x) - f(x_0)| < \varepsilon$.

We want to have

$$|f(x) - f(x_0)| = |x^2 - x_0^2| = |x - x_0||x + x_0| < \varepsilon. \tag{3.14}$$

Thus, to find an appropriate δ, we will have to find a way to bound $|x + x_0|$. Consider the interval $(x_0 - 1, x_0 + 1)$. We know that $x \to x_0$, so we can only look at values of x in this interval. This gives us

$$|x + x_0| = |x - x_0 + 2x_0| \leq |x - x_0| + |2x_0| < 1 + |2x_0|.$$

Now, choose $\delta = \frac{\varepsilon}{1+|2x_0|}$. From Equation (3.14), we then have the following:
For all x, such that $|x - x_0| < \delta$,

$$|f(x) - f(x_0)| = |x - x_0||x + x_0| < \delta|x + x_0| \leq \frac{\varepsilon}{1 + |2x_0|}(1 + |2x_0|) = \varepsilon.$$

This proves that x^2 is continuous at any $x_0 \in \mathbb{R}$.   ∎

We end this section by presenting a condition for continuity at x_0 that uses the language of convergent sequences and is equivalent to Definition 3.8.1.

> **Theorem 3.8.1** Let f be a function with domain $\mathscr{D}(f)$ and $I \subseteq \mathscr{D}(f)$ be an open interval. The function f is continuous at a point $x_0 \in I$, if and only if for every sequence $\{x_n\} \to x_0$, we have that the sequence of the function values $\{f(x_n)\}$ converges to $f(x_0)$; that is, we have $\{f(x_n)\} \to f(x_0)$.

Proof. First suppose that $f(x)$ is continuous at x_0 and $\{x_n\} \to x_0$. We will show that $\{f(x_n)\} \to f(x_0)$.

Let $f(x)$ be continuous at x_0. By definition, this means that given $\varepsilon > 0$, there is a $\delta = \delta(\varepsilon) > 0$ with the following property: if $|x_n - x_0| < \delta$, then $|f(x) - f(x_0)| < \varepsilon$. Since $\{x_n\} \to x_0$, given $\delta > 0$, there is a $N = N(\delta)$ such that $|x_n - x_0| < \delta$, for all $n > N$. But then, for $n > N$, we have $|f(x_n) - f(x_0)| < \varepsilon$. This proves that $\{f(x_n)\} \to f(x_0)$.

Next we will prove that if for every sequence with $\{x_n\} \to x_0$, we also have $\{f(x_n)\} \to f(x_0)$, then $f(x)$ is continuous at x_0. We will do a proof by contradiction. We will assume that $f(x)$ is not continuous at x_0 and find a sequence $\{x_n\} \to x_0$, for which $\{f(x_n)\}$ does not converge to $f(x_0)$, producing a contradiction.

To carry out this plan, we need to first reflect on what it means for $f(x)$ *not* to be continuous at x_0. We will do this by negating Definition 3.8.1, as we learned to do in Section 1.2.9.

Carrying out this plan, we can see that $f(x)$ is *not* continuous at x_0, provided there is a number $\varepsilon > 0$ such that for any $\delta > 0$, we can find an $x = x(\delta) \in I$, for which

$$|x - x_0| < \delta, \text{ but } |f(x) - f(x_0)| \geq \varepsilon.$$

With this in mind, we continue the proof.

Since we assumed that $f(x)$ is not continuous at x_0, there is a number $\varepsilon > 0$ such that for $\delta = \frac{1}{n}$, we can find a point $x_n \in I$, for which

$$|x_n - x_0| < \frac{1}{n}, \text{ but } |f(x_n) - f(x_0)| \geq \varepsilon. \tag{3.15}$$

We leave it as an exercise to show that we have constructed a sequence $\{x_n\}$ that converges to x_0 (see Exercise 3.39), but $\{f(x_n)\}$ does not converge to $f(x_0)$ (see Exercise 3.40). This is a contradiction, resulting from our assumption that $f(x)$ is not continuous at x_0. Therefore, $f(x)$ is continuous at x_0.

∎

Exercise 3.39 Prove that the sequence $\{x_n\}$ we constructed in the proof of Theorem 3.8.1 with the property presented in Equation (3.15) converges to x_0.

Exercise 3.40 Prove that the sequence $\{f(x_n)\}$ for the points x_n we constructed in the proof of Theorem 3.8.1 with their property presented in Equation (3.15) does not converge to $f(x_0)$.

Exercise 3.41 Let $a \neq 0$ and $b \in \mathbb{R}$. Using the definition from Exercise 3.34, prove that the function $f(x) = ax + b$ is continuous at $x = 10$. Next, prove that $f(x)$ is continuous at any point $x_0 \in \mathbb{R}$.

Exercise 3.42 Using Definition 3.8.1, prove that the function $f(x) = \sqrt{x}$ is continuous at $x = 3$. Next, show that $f(x)$ is continuous at any point x_0 in its domain $[0, \infty)$.

3.9 Summary and What to Expect Next

Our goal for this chapter was to emphasize the foundational techniques that are introduced in a beginning Real Analysis or in an Advanced Calculus course and to give you a basis for transitioning more easily from your mainly problem-based Calculus courses to a course examining the same material from a proof-oriented angle. A common beginning for such a course is the study of sequences, which gives the basis for thought at a rigorous level in analysis. This transition is usually difficult, and it may take quite some time before thinking in terms of rigorous mathematical definitions becomes a habit of mind and fosters dividends. But if you worked conscientiously through the multiple examples and exercises in this chapter, studied the definitions and proofs, and are now able to reproduce most of them with understanding, you will succeed without difficulties.

Naturally, there are vast areas of real analysis not mentioned in this chapter. In your conventional Real Analysis course you will learn more about the topology of real numbers and bounded sequences, prove important theorems regarding continuous and differentiable functions, rigorously define the Riemann Integral and study its properties, and study sequences and infinite series of functions, their convergence, and the properties of their limits. And even after you have learned all this, you would have only seen a small part of the vast field that is real analysis. Being able to see and admire its beauty is an acquired taste, and we hope to have helped you take the first steps toward developing yours.

Suggested Further Reading

We recommend the following textbooks as further reading. They are standard texts used for undergraduate courses in Real Analysis and Advanced Calculus.

Bibliography

[1] Kirkwood, James R. *An introduction to analysis*, 3rd edition. CRC Press. 2021.

[2] Royden, H. and Patrick Fitzpatrick. *Real Analysis (Classic Version)*. Pearson Modern Classics for Advanced Mathematics Series, Pearson, 2017.

4. Linear Algebra

Linear Algebra is the discipline that studies systems of linear equations, vector spaces, and linear transformations. In its simplest form, systems of linear equations allow us to find if and where two lines in the plane intersect and how to view points in the plane as two-dimensional vectors, something you have likely already seen in high school. Matrix algebra and vectors spaces help us find and study the solutions of linear systems of several variables and their properties, as well as answer questions about sets of multi-dimensional vectors of real numbers and their characteristics. Linear transformations are special kind of functions between vector spaces that preserve their essential properties. In particular, they are used to prove that every abstract vector space "generated" by a finite number of vectors behaves mathematically the same way as a space of vectors of real numbers. Further, some important questions about linear transformations can be translated into questions about vector spaces linked with matrices and can be answered using systems of linear equations.

Linear Algebra occupies a special place in the mathematics curriculum because virtually any other field of study that uses mathematics, such as statistics, computer science, economics, engineering, game development, and many others, use linear algebra extensively.

In this chapter, we introduce some fundamental properties of systems of linear equations and linear transformations using matrix theory and vector spaces.

DOI: 10.1201/9781032623849-4

4.1 Vector Spaces

A central theme of mathematics is to study a familiar structure, ascertain from the familiar structure what principles are truly important, and then create a more general structure. When you take a semester-long linear algebra course, you will prove that any general vector space behaves, mathematically speaking, like $\mathbb{R}^m$. Thus, the properties we will study here generalize directly to abstract vector spaces.

A vector space is a set V, that consists of a set of vectors, and two operations: (1) addition of vectors and (2) multiplication of a vector by a scalar (number). The operations have to satisfy certain properties (axioms). They are listed in the folowing:

Definition 4.1.1 A set V is called a *vector space* if for any vectors $\mathbf{u}, \mathbf{v}, \mathbf{w}$ in V and any scalars (numbers) c and d, the axioms below are satisfied.

Addition axioms

1. if $\mathbf{u}$ and $\mathbf{v}$ are vectors, then $\mathbf{u} + \mathbf{v}$ is a vector.
2. $\mathbf{u} + \mathbf{v} = \mathbf{v} + \mathbf{u}$.
3. $(\mathbf{u} + \mathbf{v}) + \mathbf{w} = \mathbf{u} + (\mathbf{v} + \mathbf{w})$
4. There is a vector $\mathbf{0}$ for which $\mathbf{0} + \mathbf{u} = \mathbf{u}$ for every vector $\mathbf{u}$.
5. For every vector $\mathbf{u}$, there is a vector $-\mathbf{u}$ for which $\mathbf{u} + (-\mathbf{u}) = \mathbf{0}$

Multiplication axioms

6. for each scalar c and vector $\mathbf{u}$, $c\mathbf{u}$ is a vector.
7. For scalars c and d and vector $\mathbf{u}$, $cd(\mathbf{u}) = c(d\mathbf{u})$.
8. $c(\mathbf{u} + \mathbf{v}) = c\mathbf{u} + c\mathbf{v}$.
9. For scalars c and d and vector $\mathbf{u}$, $(c + d)\mathbf{u} = c\mathbf{u} + d\mathbf{u}$.
10. $1\mathbf{u} = \mathbf{u}$.

These axioms should not be memorized. Rather, after reading them you should think "oh, that makes sense."

To develop some intuition, we start by giving an example of a vector space that you have already seen, namely the Cartesian plane in two dimensions, which we denote $\mathbb{R}^2$. Now,

$$R^2 = \{\mathbf{u} = (u_1, u_2) | u_1, u_2 \in \mathbb{R}\}.$$

Arithmetically, if $\mathbf{u} = (u_1, u_2)$ and $\mathbf{v} = (v_1, v_2)$ are vectors in $\mathbb{R}^2$, and c is a real number, then we have:

- $c\mathbf{u} = (cu_1, cu_2)$ (multiplication by a scalar);
- $\mathbf{u} + \mathbf{v} = (u_1 + v_1, u_2 + v_2)$ (addition of vectors).

To have a vector space, we also need to have axioms 4 and 5 satisfied, which we do:

- $\mathbf{0} = (0,0)$, and
- $-\mathbf{u} = (-u_1, -u_2)$.

The fact that $\mathbb{R}^2$ is a vector space should not surprise you, and the axioms in Definition 4.1.1 should look familiar. In fact, Definition 4.1.1 generalizes familiar properties of $\mathbb{R}^2$.

Among the most important definitions in linear algebra are:

Definition 4.1.2

- A *linear combination* of vectors $\mathbf{a}_1, \mathbf{a}_2, \ldots \mathbf{a}_n$ is an expression of the form where $x_1\mathbf{a}_1 + x_2\mathbf{a}_2 + \ldots x_n\mathbf{a}_n$, where $x_1, x_2, \ldots, x_n$ are scalars.
- A set of vectors $\{\mathbf{a}_1, \mathbf{a}_2, \ldots \mathbf{a}_n\}$ *spans* the vector space V if every vector in V can be written as a linear combination of $\mathbf{a}_1, \mathbf{a}_2, \ldots \mathbf{a}_n$.
- The vectors $\mathbf{a}_1, \mathbf{a}_2, \ldots \mathbf{a}_n$ are *linearly independent* if the only way $x_1\mathbf{a}_1 + x_2\mathbf{a}_2 + \ldots x_n\mathbf{a}_n = \mathbf{0}$ is when $x_1 = 0, x_2 = 0, \ldots, x_n = 0$. In this case, we also say that $\{\mathbf{a}_1, \mathbf{a}_2, \ldots, \mathbf{a}_n\}$ is a *linearly independent set*. If a set of vectors is not linearly independent, it is said to be *linearly dependent*.
- The set $\{\mathbf{a}_1, \mathbf{a}_2, \ldots, \mathbf{a}_n\}$ is a *basis* for V if it is linearly independent and it spans V.
- A function T from $\mathbb{R}^n$ to $\mathbb{R}^m$ is called a *linear transformation* if
  1. $T(\mathbf{u} + \mathbf{v}) = T(\mathbf{u}) + T(\mathbf{v})$, for all vectors $\mathbf{u}, \mathbf{v}$ in $\mathbb{R}^n$;
  2. $T(\lambda\mathbf{u}) = \lambda T(\mathbf{u})$, for all scalars λ and all vectors $\mathbf{u}$ in $\mathbb{R}^m$.

These are definitions that you will internalize with time. You may want to bookmark this page and review a definition whenever each of the terms is used, until all such terms become a part of your mathematical vocabulary.

In this chapter, you will see that vector spaces and linear transformation are closely related to matrices and systems of linear equations. Our focus will be on the vector space $\mathbb{R}^n$, but, interestingly, that will be enough, as finite-dimensional vector spaces act like $\mathbb{R}^n$ and have analogous properties. You will also see that any linear transformation from $\mathbb{R}^n$ to $\mathbb{R}^m$ can be expressed as multiplication by an $n \times m$ matrix. These ideas will be amplified in a linear algebra course, but if you keep them in mind, it will help you see the big picture of what we are doing.

4.2 Matrices and Matrix Algebra

A matrix is a rectangular array of numbers. An $m \times n$ matrix has m rows and n columns. An example of a 3×2 matrix is

$$\begin{pmatrix} 0 & -1 \\ \sqrt{2} & 6 \\ 3 & 7 \end{pmatrix}.$$

4.2.1 Basic Operations with Matrices

We want to develop an algebra for matrices. In particular, we will define how to:

1. multiply a matrix by a scalar;
2. add matrices of the same dimension, and
3. multiply matrices of compatible dimensions.

Multiplying by a scalar is intuitive. Multiply each entry in the matrix by the scalar. For example,

$$(-4)\begin{pmatrix} 2 & -1 \\ 0 & 5 \\ -3 & -2 \end{pmatrix} = \begin{pmatrix} (-4)2 & (-4)(-1) \\ (-4)0 & (-4)5 \\ (-4)(-3) & (-4)(-2) \end{pmatrix} = \begin{pmatrix} -8 & 4 \\ 0 & -20 \\ 12 & 8 \end{pmatrix}.$$

Adding matrices of the same dimension is also intuitive. Add the corresponding entries of the matrices. For example,

$$\begin{pmatrix} -3 & 1 & 2 \\ 4 & 0 & 6 \end{pmatrix} + \begin{pmatrix} 1 & -1 & 5 \\ 0 & 3 & 2 \end{pmatrix} = \begin{pmatrix} -3+1 & 1-1 & 2+5 \\ 4+0 & 0+3 & 6+2 \end{pmatrix}$$
$$= \begin{pmatrix} -2 & 0 & 7 \\ 4 & 3 & -4 \end{pmatrix}.$$

Multiplying matrices of compatible dimensions is more complicated. If we want to multiply two matrices, the matrix on the left must have the same number of entries in a row as there are entries in a column of the matrix on the right. It may be simpler to visualize this as the product of an $(m \times k)$ matrix on the left with an $(k \times n)$ matrix on the right will be a $(m \times n)$ matrix. Also an $(n \times k)$ matrix multiplied by an $(m \times p)$ matrix is not defined unless $k = m$.

The mechanics of matrix multiplication are more complex than the previous arithmetic operations, but it is constructed so that the composition of linear transformations (a later topic) can be represented as the product of matrices. We describe matrix multiplication in stages.

Case 1: Multiplying a matrix consisting of one row by a matrix consisting of one column. Here the row matrix will be on the left and the column matrix on the right. Also, the row matrix and column matrix must have the same number of entries. The definition of this process is

$$\begin{pmatrix} a_1 & a_2 & \cdots & a_n \end{pmatrix} \begin{pmatrix} b_1 \\ b_2 \\ \vdots \\ b_n \end{pmatrix} = \begin{pmatrix} a_1b_1 + a_2b_2 + \cdots + a_nb_n \end{pmatrix}.$$

Note that a $(1 \times n)$ matrix multiplied by a $(n \times 1)$ matrix is a (1×1) matrix; i.e., a number.

Case 2: Multiplying a matrix consisting of one row on the left by a matrix consisting of more than one column on the right. Here again, the number of entries in the row must equal the number of entries in each column. An example is

$$\begin{pmatrix} a_1 & a_2 & \cdots & a_n \end{pmatrix} \begin{pmatrix} b_1 & c_1 \\ b_2 & c_2 \\ \vdots & \\ b_n & c_n \end{pmatrix}$$
$$= \begin{pmatrix} a_1b_1 + a_2b_2 + \cdots + a_nb_n & a_1c_1 + a_2c_2 + \cdots + a_nc_n \end{pmatrix}.$$

So a $(1 \times n)$ matrix on the left multiplied by an $(n \times 2)$ on the right, gives a (1×2) matrix.

Case 3: Multiplying a matrix of more than one row by a matrix of one column where the column matrix is on the right. In this case, we multiply each row by the column, and put the results in a column. Two examples of this are

$$\begin{pmatrix} a_1 & a_2 & \cdots & a_n \\ b_1 & b_2 & \cdots & b_n \end{pmatrix} \begin{pmatrix} c_1 \\ c_2 \\ \vdots \\ c_n \end{pmatrix} = \begin{pmatrix} a_1c_1 + a_2c_2 + \cdots + a_nc_n \\ b_1c_1 + b_2c_2 + \cdots + b_nc_n \end{pmatrix}.$$

So a $(2 \times n)$ matrix on the left multiplied by an $(n \times 1)$ matrix on the right, gives a (2×1) matrix.

An example for multiplying a $(3 \times n)$ matrix on the left by a $(n \times 1)$ matrix on the right is done as follows:

$$
\begin{pmatrix} a_1 & a_2 & \cdots & a_n \\ b_1 & b_2 & \cdots & b_n \\ c_1 & c_2 & \cdots & c_n \end{pmatrix} \begin{pmatrix} d_1 \\ d_2 \\ \vdots \\ d_n \end{pmatrix} = \begin{pmatrix} a_1 d_1 + a_2 d_2 + \cdots + a_n d_n \\ b_1 d_1 + b_2 d_2 + \cdots + b_n d_n \\ c_1 d_1 + c_2 d_2 + \cdots + c_n d_n \end{pmatrix}.
$$

Case 4: Multiplying any two *compatible* matrices. Again, in order to be compatible for multiplication, the matrix on the left must have the same number of entries in a row as there are entries in a column of the matrix on the right. One example is:

$$
\begin{pmatrix} a_1 & a_2 & a_3 \\ b_1 & b_2 & b_3 \end{pmatrix} \begin{pmatrix} c_1 & c_2 \\ d_1 & d_2 \\ e_1 & e_2 \end{pmatrix} = \begin{pmatrix} a_1 c_1 + a_2 d_1 + a_3 e_1 & a_1 c_2 + a_2 d_2 + a_3 e_2 \\ b_1 c_1 + b_2 d_1 + b_3 e_1 & b_1 c_2 + b_2 d_2 + b_3 e_2 \end{pmatrix}.
$$

So, a

(2×3) matrix multiplied by a (3×2) matrix gives a (2×2) matrix.

In general, if A is a $m \times k$ matrix and B is a $k \times n$ matrix, say

$$
A = \begin{pmatrix} a_{11} & a_{12} & \cdots & a_{1k} \\ a_{21} & a_{22} & \cdots & a_{2k} \\ \vdots & \vdots & & \vdots \\ a_{m1} & a_{m2} & \cdots & a_{1k} \end{pmatrix} \quad \text{and} \quad B = \begin{pmatrix} b_{11} & b_{12} & \cdots & a_{1n} \\ a_{21} & a_{22} & \cdots & a_{2n} \\ \vdots & \vdots & & \vdots \\ a_{k1} & a_{k2} & \cdots & a_{kn} \end{pmatrix},
$$

then the product matrix AB has dimensions $m \times n$ and the formula for the ij-th entry (the element in the i-th row and j-th column) of the matrix AB is

$$
(AB)_{ij} = \sum_{l=1}^{k} a_{il} b_{lj}.
$$

Exercise 4.1 Are the following multiplications possible? If so, find the resulting matrix. If not, explain why.

1. $(0 \quad 5) \begin{pmatrix} 2 \\ -1 \\ 0 \\ 8 \end{pmatrix}$;

2. $\begin{pmatrix} 1 & 1 & 0 \\ 2 & 3 & -1 \\ 4 & 2 & 2 \end{pmatrix} \begin{pmatrix} 2 & 3 & -1 \\ 20 & 3 & 0 \\ 5 & 1 & 2 \end{pmatrix}$;

3. $\begin{pmatrix} 2 & 3 & -1 \\ 20 & 3 & 0 \\ 5 & 1 & 2 \end{pmatrix} \begin{pmatrix} 1 & 1 & 0 \\ 2 & 3 & -1 \\ 4 & 2 & 2 \end{pmatrix}$.

∎

Exercise 4.2 Suppose

$$A = \begin{pmatrix} 2 & 0 \\ 3 & 1 \end{pmatrix}, \quad B = \begin{pmatrix} -1 & 2 \\ -3 & 2 \\ 1 & 4 \end{pmatrix}, \quad C = \begin{pmatrix} 1 & 5 & 2 \\ -3 & 0 & 6 \end{pmatrix},$$

$$D = \begin{pmatrix} 4 & -2 & 2 \\ 0 & 1 & 5 \\ 2 & 0 & -3 \end{pmatrix}.$$

Where the computations are possible, compute
1. BC;
2. $CB + 4D$;
3. BA;
4. $2A + 4C$.
Where a computation is not possible, explain why.

∎

.2 Other Properties of Matrices

Here we list some important properties of the matrix operations we introduced in the previous section. We give them in the form of a theorem, which you will prove in your conventional linear algebra class. These properties will likely feel familiar to you, as they mirror some well-known properties of real numbers. However, the note we make following this theorem points out some important distinctions you need to be aware of.

Theorem 4.2.1 If the sizes of the matrices A, B, and C are such that the operations below can be performed, and α and β are scalars, then
1. $A + B = B + A$;
2. $(A + B) + C = A + (B + C)$;

3. $(AB)C = A(BC)$;
4. $A(B+C) = AB+AC$;
5. $(B+C)A = BA+CA$;
6. $\alpha(A+B) = \alpha A + \alpha B$;
7. $(\alpha+\beta)A = \alpha A + \beta A$;
8. $\alpha(\beta A) = (\alpha\beta)A$;
9. $\alpha(AB) = (\alpha A)B = A(\alpha B)$.

There are matrices that have properties that are analogous to those of numbers. The matrix all of whose entries are 0 is called the zero matrix and is denoted $\mathbf{0}$. It has the property that $A + \mathbf{0} = A$, provided the dimensions are compatible.

Also, the $n \times n$ matrix that has 1's on the main diagonal and 0's elsewhere

$$I_n = \begin{pmatrix} 1 & 0 & 0 & \cdots & 0 \\ 0 & 1 & 0 & \cdots & 0 \\ 0 & 0 & 1 & \cdots & 0 \\ & & & \ddots & \\ 0 & 0 & 0 & \cdots & 1 \end{pmatrix}$$

multiplies like the number 1. That is, if A is an $n \times n$ matrix, then

$$AI_n = I_nA = A.$$

The matrix I_n is the *identity matrix* of size n.

It is very important to be aware of the following facts that do *not* mimic the properties of real numbers:

1. Even if A and B are square matrices of the same dimension, it is not necessarily true that $AB = BA$;
2. It is possible that A and B are square matrices of the same dimension with $AB = \mathbf{0}$ but neither A not B is the zero matrix.
3. It is also possible to have $AC = BC$ but $A \neq B$, even if C is not a zero matrix.

We provide some examples next.

■ **Example 4.1** To illustrate cases 1 and 2 above, let

$$A = \begin{pmatrix} 1 & 0 \\ 0 & 0 \end{pmatrix}, \quad B = \begin{pmatrix} 0 & 0 \\ 1 & 1 \end{pmatrix}.$$

Then,

$$AB = \begin{pmatrix} 1 & 0 \\ 0 & 0 \end{pmatrix} \begin{pmatrix} 0 & 0 \\ 1 & 1 \end{pmatrix} = \begin{pmatrix} 0 & 0 \\ 0 & 0 \end{pmatrix}, \text{ but } BA = \begin{pmatrix} 0 & 0 \\ 1 & 1 \end{pmatrix} \begin{pmatrix} 1 & 0 \\ 0 & 0 \end{pmatrix} = \begin{pmatrix} 0 & 0 \\ 1 & 0 \end{pmatrix}.$$

■

■ **Example 4.2** To illustrate case 3, let

$$A = \begin{pmatrix} 1 & 0 \\ 0 & 1 \end{pmatrix}, \quad B = \begin{pmatrix} 1 & 0 \\ 1 & 1 \end{pmatrix}, \quad C = \begin{pmatrix} 0 & 0 \\ 1 & 0 \end{pmatrix}.$$

Then

$$AC = \begin{pmatrix} 1 & 0 \\ 0 & 1 \end{pmatrix} \begin{pmatrix} 0 & 0 \\ 1 & 0 \end{pmatrix} = \begin{pmatrix} 0 & 0 \\ 1 & 0 \end{pmatrix}, \quad BC = \begin{pmatrix} 1 & 0 \\ 1 & 1 \end{pmatrix} \begin{pmatrix} 0 & 0 \\ 1 & 0 \end{pmatrix} = \begin{pmatrix} 0 & 0 \\ 1 & 0 \end{pmatrix},$$

the matrix C is not the zero matrix, and yet $A \neq B$.

■

3 Two-by-Two Matrices – Determinants and Inverses

We now consider the special case of 2×2 matrices to illustrate the concept of an inverse matrix, which will be covered in its generality in a standard Linear Algebra course for square matrices of any size.

Consider the 2×2 matrix A,

$$A = \begin{pmatrix} a & b \\ c & d \end{pmatrix}, \text{ where } a, b, c, d \in \mathbb{R}. \tag{4.1}$$

and the 2×2 identity matrix I,

$$I = \begin{pmatrix} 1 & 0 \\ 0 & 1 \end{pmatrix}.$$

Definition 4.2.1 Given a square matrix A, the matrix B with the property

$$AB = BA = I$$

is said to be the *inverse matrix* of A and is denoted by $B = A^{-1}$ (this is read as "A-inverse").

This definition applies to any $n \times n$ matrices, but here we will focus only on the special case $n = 2$, in which the computations are much simpler.

■ **Example 4.3** Consider

$$A = \begin{pmatrix} 4 & 3 \\ 3 & 2 \end{pmatrix} \text{ and } B = \begin{pmatrix} -2 & 3 \\ 3 & -4 \end{pmatrix}.$$

We will show that $B = A^{-1}$ by checking that $AB = BA = I$.

$$AB = \begin{pmatrix} 4 & 3 \\ 3 & 2 \end{pmatrix} \begin{pmatrix} -2 & 3 \\ 3 & -4 \end{pmatrix} = \begin{pmatrix} 1 & 0 \\ 0 & 1 \end{pmatrix} \quad \text{and}$$

$$BA = \begin{pmatrix} -2 & 3 \\ 3 & -4 \end{pmatrix} \begin{pmatrix} 4 & 3 \\ 3 & 2 \end{pmatrix} = \begin{pmatrix} 1 & 0 \\ 0 & 1 \end{pmatrix}.$$

This establishes that $B = A^{-1}$.   ■

Not every matrix A has an inverse, and it is convenient to have a condition that would allow us to determine the class of matrices for which the inverse exists. Also, if the inverse exists, we would want to know how to find it. The answer to these questions depends on a special number that can be computed for each square matrix called the determinant of the matrix. In the case of 2×2 matrices, we have a simple formula. Finding the determinant of a square matrix in the general cases is more technical, and we will not discuss it here.

Definition 4.2.2 The *determinant* det(A) of a 2×2 matrix

$$A = \begin{pmatrix} a & b \\ c & d \end{pmatrix},$$

is the real number computed as

$$\det(A) = ad - bc. \tag{4.2}$$

■ **Example 4.4** Consider the matrices A and B from Example 4.3. We have

$$\det(A) = (4)(2) - (3)(3) = -1, \quad \text{and} \quad \det(B) = (-2)(-4) - (3)(3) = -1.$$

                                                                        ■

Knowing how to calculate the determinant of a 2×2 matrix A, we can now formulate a condition for the existence of A^{-1} and how to find it.

Theorem 4.2.2 Given a 2×2 matrix

$$A = \begin{pmatrix} a & b \\ c & d \end{pmatrix},$$

the inverse matrix A^{-1} can be found as

$$A^{-1} = \frac{1}{\det(A)} \begin{pmatrix} d & -b \\ -c & a \end{pmatrix}, \quad \text{for} \quad \det(A) \neq 0. \tag{4.3}$$

When $\det A = 0$, an inverse matrix A^{-1} does not exist.

Proof. We will show that the matrix from Equation (4.3) satisfies the condition from Definition 4.2.1.

We have

$$A^{-1}A = \frac{1}{\det(A)} \begin{pmatrix} d & -b \\ -c & a \end{pmatrix} \begin{pmatrix} a & b \\ c & d \end{pmatrix} = \frac{1}{ad - bc} \begin{pmatrix} da - bc & 0 \\ 0 & -cb + ad \end{pmatrix}$$

$$= \begin{pmatrix} 1 & 0 \\ 0 & 1 \end{pmatrix}.$$

Checking that $AA^{-1} = I$ is done similarly. ∎

■ **Example 4.5** Let's apply Theorem 4.2.2 to find the inverse matrix of the matrix A from Example 4.3. We have $\det(A) = (4)(2) - (3)(3) = -1$, and, using Equation (4.3) gives us

$$A^{-1} = \frac{1}{-1} \begin{pmatrix} 2 & -3 \\ -3 & 4 \end{pmatrix} = \begin{pmatrix} -2 & 3 \\ 3 & -4 \end{pmatrix},$$

just as we established in Example 4.3. ■

■ **Example 4.6** Consider the matrices

$$A = \begin{pmatrix} 3 & 1 \\ -3 & -2 \end{pmatrix}, B = \begin{pmatrix} 2 & 3 \\ 6 & 9 \end{pmatrix}, \quad \text{and} \quad C = \begin{pmatrix} 0 & 0 \\ 0 & 0 \end{pmatrix}.$$

For each matrix, determine if the inverse exists. If it does, find the inverse matrix.

For A, we have $\det(A) = (3)(-2) - (1)(-3) = -6 + 3 = -3 \neq 0$. Thus, A^{-1} exists. From Equation (4.3), we find

$$A^{-1} = \frac{1}{-3} \begin{pmatrix} 3 & 1 \\ -3 & -2 \end{pmatrix} = \begin{pmatrix} -1 & -\frac{1}{3} \\ 1 & \frac{2}{3} \end{pmatrix}.$$

For B, we have $\det(B) = (2)(9) - (3)(6) = 0$, so the matrix B does not have an inverse. Similarly, for the matrix C, we have $\det(C) = 0$, so C^{-1} does not exist. Note that the matrix B gives an example of a nonzero matrix that does not have an inverse. ■

Exercise 4.3 For each of the matrices below:
1. Compute the determinant;
2. Determine if the matrix has an inverse;
3. If an inverse exists, find the inverse matrix using Theorem 4.2.2;
4. Verify that the inverse satisfies the conditions in Definition 4.2.1.

$$A = \begin{pmatrix} 7 & -2 \\ -5 & 2 \end{pmatrix}, \ B = \begin{pmatrix} 6 & 12 \\ 1 & 2 \end{pmatrix}, \ C = \begin{pmatrix} 1 & 0 \\ 0 & 5 \end{pmatrix}, \ \text{and} \ D = \begin{pmatrix} 0 & 1 \\ 0 & 5 \end{pmatrix}.$$

4.3 Systems of Linear Equations

Definition 4.3.1 A linear equation in the variables $x_1, \ldots, x_n$ is an expression of the form

$$a_1 x_1 + \cdots + a_n x_n = b$$

where $a_1, \ldots, a_n$, and b are constants. A solution to the equation $a_1 x_1 + \cdots + a_n x_n = b$ is an ordered n-tuple of numbers $(s_1, \ldots, s_n)$ for which $a_1 s_1 + \cdots + a_n s_n = b$.

A *system of linear equations* in the variables $x_1, \ldots, x_n$ is a finite set of linear equations of the form

$$
\begin{array}{ccccccc}
a_{11}x_1 & + & a_{12}x_2 & + \ldots & a_{1m}x_m & = & b_1 \\
a_{21}x_1 & + & a_{22}x_2 & + \ldots & a_{2m}x_m & = & b_2 \\
& \vdots & & & & & \\
a_{n1}x_1 & + & a_{12}x_2 & + \ldots & a_{1m}x_m & = & b_m
\end{array}
$$

A solution to a *system of linear equations* is an ordered n-tuple of numbers $(s_1, s_2, \ldots, s_n)$ which is a solution to every equation in the system. A system of linear equations has either

- no solution;
- exactly one solution, or
- infinitely many solutions.

The geometric explanation for this in two dimensions is that, in two variables, a single linear equation is $y = ax + b$, and its graph is a line. The graphs of two lines in the plane are either

- parallel, so we have no solution;
- intersecting, in which case there is exactly one solution, or
- identical, where we have infinitely many solutions.

You probably learned how to solve simple linear systems in your earlier courses. We will usually use computers to assist us with solving systems of linear equations. This involves transforming the system into what is called *row reduced form*.

To illustrate the main idea, we will now take a simple system of linear equations and solve it in two ways. In the left hand column we solve it by hand and, in the right column, we mimic what the computer does when the system is represented by a matrix. Notice that in the matrix version, the first column gives the coefficients for the first variable (x), the second column corresponds to the second variable (y), and the third column corresponds to the right-hand side of the system of equations. As long as we know the order of variables, there is no need to explicitly include them when we use the matrix form.

Consider now the following system of linear equations.

$$\begin{aligned} 2x + 3y &= 5 \\ 3x + 4y &= 9 \end{aligned}$$

For this system, the *coefficient matrix* is $\begin{pmatrix} 2 & 3 \\ 3 & 4 \end{pmatrix}$ and the matrix $\begin{pmatrix} 2 & 3 & 5 \\ 3 & 4 & 9 \end{pmatrix}$ is the *augmented matrix* of the system, where the coefficient matrix is augmented to include in its last column the right-hand side values of the system.

We now solve the system of equations and show how each step is reflected on the augmented matrix.

Start with the original system and its augmented matrix:

$$\begin{aligned} 2x + 3y &= 5 \\ 3x + 4y &= 9 \end{aligned} \qquad \begin{pmatrix} 2 & 3 & 5 \\ 3 & 4 & 9 \end{pmatrix}.$$

Divide the first equation by 2 and the second equation by 3.

$$\begin{aligned} x + \tfrac{3}{2}y &= \tfrac{5}{2} \\ x + \tfrac{4}{3}y &= 3 \end{aligned} \qquad \begin{pmatrix} 1 & \tfrac{3}{2} & \tfrac{5}{2} \\ 1 & \tfrac{4}{3} & 3 \end{pmatrix}.$$

Subtract the second equation from the first.

$$\begin{aligned} x + \tfrac{3}{2}y &= \tfrac{5}{2} \\ \tfrac{1}{6}y &= -\tfrac{1}{2} \end{aligned} \qquad \begin{pmatrix} 1 & \tfrac{3}{2} & \tfrac{5}{2} \\ 0 & \tfrac{1}{6} & -\tfrac{1}{2} \end{pmatrix}.$$

Multiply the second equation by 6.

$$\begin{aligned} x + \tfrac{3}{2}y &= \tfrac{5}{2} \\ y &= -3 \end{aligned} \qquad \begin{pmatrix} 1 & \tfrac{3}{2} & \tfrac{5}{2} \\ 0 & 1 & -3 \end{pmatrix}.$$

Multiply the second equation by $-\frac{3}{2}$ and add it to the first equation to get

$$\begin{array}{rcl} x & = & 7 \\ y & = & -3 \end{array} \qquad \begin{pmatrix} 1 & 0 & 7 \\ 0 & 1 & -3 \end{pmatrix}.$$

The matrix $\begin{pmatrix} 1 & 0 & 7 \\ 0 & 1 & -3 \end{pmatrix}$ is an example of a matrix that is in row reduced form, and it provides an easy way to give the solution. In general, the row reduced form is the simplest form from which to interpret the solution for a system of linear equations. As long as you understand how to do that (as in the example above), there is no need to find it "by hand." Many calculators and computational systems have that capability.

> The free web-based Desmos Matrix Calculator `https://www.desmos.com/matrix` is a good tool for finding the row reduced form of a matrix. It is intuitive and easy to use.

We will return to this topic again when we consider more complex systems of equations. For now, here is how we recognize if a matrix is in row reduced form

Definition 4.3.2 A matrix is in *row reduced form* (sometimes also called *row reduced echelon form*) when it has the following characteristics:
1. Reading from left to right, the first nonzero entry in each row is 1. It is called a *leading 1*;
2. A leading 1 is the only nonzero entry in its column;
3. A leading 1 in a row is always strictly to the right of the leading 1 in the row above it;
4. Any rows that are all 0 will appear as the bottom rows in row reduced form.

The variable associated with the column where a leading 1 appears is called a *leading variable*. The variables that do not correspond to a column with a leading 1 are called *free variables*.

■ **Example 4.7** The next two matrices are in row reduced form.

1. $\begin{pmatrix} 0 & 1 & 3 & 0 & 2 & 0 & 6 \\ 0 & 0 & 0 & 1 & 0 & 5 & 3 \\ 0 & 0 & 0 & 0 & 0 & 0 & 0 \end{pmatrix}$.

If the variables for this matrix are $x_1, x_2, x_3, x_4, x_5, x_6$ and x_7, then the leading variables are x_2 and x_4. The variables x_1, x_3, x_5, x_6, and x_7 are free variables.

2. $\begin{pmatrix} 1 & 0 & 0 & 5 & 0 \\ 0 & 1 & 0 & 4 & 0 \\ 0 & 0 & 1 & 6 & 0 \\ 0 & 0 & 0 & 0 & 0 \end{pmatrix}.$

If the variables for this matrix are x_1, x_2, x_3, x_4, and x_5, then the leading variables are x_1, x_2, and x_3. The variables x_4 and x_5 are free variables.

∎

■ **Example 4.8** The following matrices are not in row reduced form.

1. $\begin{pmatrix} 0 & 1 & 2 & 0 & 3 \\ 1 & 0 & 0 & 0 & 9 \\ 0 & 0 & 0 & 1 & 5 \end{pmatrix};$

2. $\begin{pmatrix} 1 & 1 & 0 & 0 & 3 \\ 0 & 0 & 2 & 0 & 9 \\ 0 & 0 & 0 & 1 & 5 \end{pmatrix};$

3. $\begin{pmatrix} 1 & 0 & 0 & 5 & 0 \\ 0 & 1 & 2 & 4 & 0 \\ 0 & 0 & 1 & 6 & 0 \\ 0 & 0 & 0 & 0 & 0 \end{pmatrix}.$

For matrix 1, condition 3 of Definition 4.3.2 is violated. The second row's leading 1 is to the left of the 1 in the first row.

For matrix 2, condition 1 of Definition 4.3.2 is violated, as the first nonzero entry in the second row is not 1.

For matrix 3, the third row's leading 1 is not the only nonzero entry in its column. So condition 2 in Definition 4.3.2 is not satisfied. ∎

Exercise 4.4 For each of the matrices, determine if the matrix is in row reduced form. Where that's not the case, explain which condition or conditions in Definition 4.3.2 are not satisfied.

1. $\begin{pmatrix} 1 & 0 & 3 & 1 \\ 0 & 1 & 0 & 0 \\ 0 & 0 & 1 & 2 \end{pmatrix};$

2. $\begin{pmatrix} 1 & 0 & 2 & 5 \\ 0 & 1 & 1 & 0 \\ 0 & 0 & 0 & 1 \end{pmatrix};$

3. $\begin{pmatrix} 1 & 0 & 1 & 1 \\ 0 & 1 & -3 & 0 \\ 0 & 0 & 0 & 0 \end{pmatrix}$;

4. $\begin{pmatrix} 1 & 0 & 1 & 1 \\ 0 & 1 & 7 & 0 \\ 2 & 0 & 0 & 1 \\ 0 & 1 & 0 & 4 \end{pmatrix}$;

5. $\begin{pmatrix} 0 & 0 & 1 & 1 \\ 0 & 0 & 0 & 1 \\ 0 & 0 & 0 & 1 \\ 0 & 0 & 0 & 0 \end{pmatrix}$;

6. $\begin{pmatrix} 0 & 0 & 1 & 6 \\ 0 & 0 & 0 & 1 \\ 0 & 0 & 0 & 0 \\ 0 & 0 & 0 & 0 \end{pmatrix}$.

4.3.1 General Solution of a System of Linear Equations

We already noted that any system of linear equations either has a unique solution, infinitely many solutions or no solution at all. Here we will see how to determine this for a given system of linear equations, using the row reduced form of its augmented matrix. We will do this by considering an example for each case first.

■ **Example 4.9 – A system with a unique solution.**

Consider the system of equations

$$
\begin{aligned}
4x_1 &+ 3x_2 &+ 3x_3 &= 14 \\
-2x_1 &+ x_2 &- 4x_3 &= -2 \\
2x_1 &+ 4x_2 &- 6x_3 &= 1
\end{aligned}
$$

Construct the augmented matrix to obtain

$$
\begin{pmatrix} 4 & 3 & 3 & 14 \\ -2 & 1 & -4 & -2 \\ 2 & 4 & -6 & 1 \end{pmatrix}.
$$

Now use the Desmos or another matrix calculator to find the row reduced form of this matrix. We obtain

$$\begin{pmatrix} 1 & 0 & 0 & -1.3 \\ 0 & 1 & 0 & 4.2 \\ 0 & 0 & 1 & 2.2 \end{pmatrix}.$$

Writing this back as a system, gives the unique solution

$$\begin{aligned} x_1 & = & -1.3 \\ x_2 & = & 4.2 \\ x_3 & = & 2.2 \end{aligned}$$

∎

Notice the following: (1) the number of variables in this example is the same as the number of equations (that is, the coefficient matrix is a square matrix) and (2) that in the row reduced form of the augmented matrix, every column has a leading 1. When those conditions are satisfied, the system will always have a unique solution.

What happens when the number of variables is different than the number of equations? In that case, we either have no solutions or infinitely many solutions.

■ **Example 4.10 – A system with no solutions**

Consider the system

$$\begin{aligned} x_1 & - & 2x_2 & - & x_3 & + & 3x_4 & = & 0 \\ -2x_1 & + & 4x_2 & + & 5x_3 & - & 5x_4 & = & 3 \\ 3x_1 & - & 6x_2 & - & 6x_3 & + & 8x_4 & = & 2 \end{aligned}$$

The augmented matrix of this system is

$$\begin{pmatrix} 1 & -2 & -1 & 3 & 0 \\ -2 & 4 & 5 & -5 & 3 \\ 3 & -6 & -6 & 8 & 2 \end{pmatrix}$$

Using Desmos or another matrix calculator, we obtain the row reduced form of this matrix to be

$$\begin{pmatrix} 1 & -2 & 0 & \frac{10}{3} & 0 \\ 0 & 0 & 1 & \frac{1}{3} & 0 \\ 0 & 0 & 0 & 0 & 1 \end{pmatrix}$$

Now let's write this back as a system of equations:

$$\begin{array}{rcrcrcrcl}
x_1 & - & 2x_2 & + & 0x_3 & + & \frac{10}{3}x_4 & = & 0 \\
0x_1 & + & 0x_2 & + & x_3 & + & \frac{1}{3}x_4 & = & 0 \\
0x_1 & + & 0x_2 & + & 0x_3 & + & 0x_4 & = & 1
\end{array}$$

Of course, in previous examples, we never wrote the terms in the system with zero coefficients and we won't do it ever again. In this case, though, we wanted to make clear that the last row of the augmented matrix corresponds to an equation, which has no solution – no matter what values for the x's we pick, the left side of the third equation will be zero, while the right side is one. Thus, the system has no solution.   ∎

The last example shows how to recognize, in general, when a system doesn't have a solution: this happens when the row reduced form of the augmented matrix contains a row in which all entries are zeros except for the entry in the last column, which is *not* a zero.

■ **Example 4.11 – A system with infinitely many solutions**
Consider the system

$$\begin{array}{rcrcrcrcrcl}
3x_1 & - & 2x_3 & + & x_4 & + & 6x_5 & - & x_6 & = & 3 \\
x_2 & - & \frac{4}{15}x_3 & - & \frac{2}{3}x_4 & + & \frac{1}{5}x_5 & + & \frac{7}{16}x_6 & = & \frac{9}{5}
\end{array}$$

The augmented matrix of this system has the following row reduced form:

$$\begin{pmatrix}
1 & 0 & -\frac{2}{3} & \frac{1}{3} & 2 & -\frac{1}{3} & 1 \\
0 & 1 & -\frac{4}{15} & -\frac{2}{3} & \frac{1}{5} & \frac{7}{16} & \frac{9}{5}
\end{pmatrix}.$$

From here, we see that x_1 and x_2 are leading variables, and x_3, x_4, x_5, and x_6 are free variables. Let's use the row reduced form of the augmented matrix to go back to the form where the variables are expressed explicitly:

$$\begin{array}{rcrcrcrcrcl}
x_1 & - & \frac{2}{3}x_3 & + & \frac{1}{3}x_4 & + & 2x_5 & - & \frac{1}{3}x_6 & = & 1 \\
x_2 & - & \frac{4}{15}x_3 & - & \frac{2}{3}x_4 & + & \frac{1}{5}x_5 & + & \frac{7}{16}x_6 & = & \frac{9}{5}
\end{array}$$

One then expresses the leading variables in terms of the free variables. In this case we have:

$$\begin{array}{rclcrcrcrcl}
x_1 & = & \frac{2}{3}x_3 & - & \frac{1}{3}x_4 & - & 2x_5 & + & \frac{1}{3}x_6 & + & 1 \\
x_2 & = & \frac{4}{15}x_3 & + & \frac{2}{3}x_4 & - & \frac{1}{5}x_5 & - & \frac{7}{16}x_6 & + & \frac{9}{5}
\end{array}$$

Picking any arbitrary values for the free variables and calculating the values of the leading variables from the above equations, gives a solution for the system. Since we can pick values for the free variables in infinitely many ways, the system has infinitely many solutions of the form above. So, choosing $x_3 = s$, $x_4 = t$, $x_5 = u$, and $x_6 = v$, where s, t, u, v are arbitrary numbers, gives the *solution set* of the system:

$$
\begin{aligned}
x_1 &= \tfrac{2}{3}s - \tfrac{1}{3}t - 2u + \tfrac{1}{3}v + 1 \\
x_2 &= \tfrac{4}{15}s + \tfrac{2}{3}t - \tfrac{1}{5}u - \tfrac{7}{16}v + \tfrac{9}{5} \\
x_3 &= s \\
x_4 &= t \\
x_5 &= u \\
x_6 &= v
\end{aligned}
$$

Choosing any values for s, t, u, v and substituting in the equations above, gives a solution for the system. ∎

We can readily generalize the last example. It shows that systems of linear equations with at least one free variable, have infinitely many solutions.

Exercise 4.5 For each of the systems below, give the augmented form of the matrix and find the row reduced form of the matrix. If a solution exists, give the solution set, and specify whether the system has a single solution or infinitely many solutions.

1.
$$
\begin{aligned}
2x - 3y &= 6 \\
x + 5y &= 9
\end{aligned}
$$

2.
$$
\begin{aligned}
x - 2y + 3z &= 5 \\
5x - y + 2z &= 4
\end{aligned}
$$

3.
$$
\begin{aligned}
3x_1 - 5x_2 + x_3 - 2x_4 &= 0 \\
6x_1 - 10x_2 + 2x_3 - 4x_4 &= -6
\end{aligned}
$$

4.
$$
\begin{aligned}
4x_1 + 3x_2 + 3x_3 &= 14 \\
-2x_1 + x_2 - 4x_3 &= -2 \\
2x_1 + 4x_2 - 6x_3 &= 1
\end{aligned}
$$

$$
\begin{array}{rrrrrr}
& x_1 & + & & + & 2x_3 & = & 4 \\
5. & -2x_1 & + & x_2 & - & 4x_3 & = & -2 \\
& & & 4x_2 & - & 6x_3 & = & 1
\end{array}
$$

$$
\begin{array}{rrrrrr}
& x_1 & - & 2x_2 & - & x_3 & + & 3x_4 & = & 0 \\
6. & -2x_1 & + & 4x_2 & + & 5x_3 & - & 5x_4 & = & 3 \\
& 3x_1 & - & 6x_2 & - & 6x_3 & + & 8x_4 & = & 2
\end{array}
$$

4.3.2 Matrix Form and Vector Form of a System of Linear Equations

In the previous section we discussed matrices and how to use them to solve systems of linear equations. Here, we show how a system of equations can be written in matrix form, as well as in vector form. This makes the notation less cumbersome and, more importantly, provides intuitive understanding of some important definitions that follow.

■ **Example 4.12** Consider the system of equations

$$
\begin{array}{rrrrrr}
3x_1 & + & 7x_2 & - & x_3 & = & 11 \\
x_1 & + & 2x_2 & - & x_3 & = & 3 \\
2x_1 & + & 4x_2 & - & 2x_3 & = & 10
\end{array}
$$

Denote by A the coefficient matrix, by $\mathbf{x}$ the vector of unknowns, and by $\mathbf{b}$ the vector of the numbers on the right. That is,

$$
A = \begin{pmatrix} 3 & 7 & -1 \\ 1 & 2 & -1 \\ 2 & 4 & -2 \end{pmatrix}, \qquad \mathbf{x} = \begin{pmatrix} x_1 \\ x_2 \\ x_3 \end{pmatrix}, \qquad \mathbf{b} = \begin{pmatrix} 11 \\ 3 \\ 10 \end{pmatrix}.
$$

Using matrix multiplication, verify that the system of equations can be written in *matrix form* as

$$
\begin{pmatrix} 3 & 7 & -1 \\ 1 & 2 & -1 \\ 2 & 4 & -2 \end{pmatrix} \begin{pmatrix} x_1 \\ x_2 \\ x_3 \end{pmatrix} = \begin{pmatrix} 11 \\ 3 \\ 10 \end{pmatrix},
$$

or, with the notation we have introduced, as $A\mathbf{x} = \mathbf{b}$.

Further, denote the columns of A by $\mathbf{a}_1, \mathbf{a}_2$, and $\mathbf{a}_3$. That is,

$$
\mathbf{a}_1 = \begin{pmatrix} 3 \\ 1 \\ 2 \end{pmatrix}, \qquad \mathbf{a}_2 = \begin{pmatrix} 7 \\ 2 \\ 4 \end{pmatrix}, \qquad \mathbf{a}_3 = \begin{pmatrix} -1 \\ -1 \\ -2 \end{pmatrix}.
$$

Now the system of equations can be written in the following *vector form*:

$$x_1 \begin{pmatrix} 3 \\ 1 \\ 2 \end{pmatrix} + x_2 \begin{pmatrix} 7 \\ 2 \\ 4 \end{pmatrix} + x_3 \begin{pmatrix} -1 \\ -1 \\ -2 \end{pmatrix} = \begin{pmatrix} 11 \\ 3 \\ 10 \end{pmatrix},$$

or, equivalently, we have *the vector form*

$$x_1 \mathbf{a}_1 + x_2 \mathbf{a}_2 + x_3 \mathbf{a}_3 = \mathbf{b}.$$

■

Notice that in Example 4.12 above, the number of equations was equal to the number of unknowns, which, as we have already seen in earlier examples, is not always the case. In general, if we have m equations and n unknowns, the coefficient matrix A is of size $m \times n$, $\mathbf{x}$ is in $\mathbb{R}^n$, and $\mathbf{b}$ is in $\mathbb{R}^m$. The nice thing, as our next example shows, is that as long as the number of variables n does not change, the matrix and vector forms of the system remain the same.

■ **Example 4.13** Consider the system of equations

$$\begin{array}{ccccccc} x_1 & + & x_2 & + & 2x_3 & = & 11 \\ 2x_1 & + & x_2 & + & 3x_3 & = & 3 \\ 3x_1 & + & x_2 & + & 4x_3 & = & 10 \\ 4x_1 & + & x_2 & + & 5x_3 & = & -1 \end{array}$$

Here, we have $m = 4$ equations and $n = 3$ unknowns. The coefficient matrix, the vector $\mathbf{x}$ of the unknowns, and the vector $\mathbf{b}$ formed from the right-hand side numbers are:

$$A = \begin{pmatrix} 1 & 1 & 2 \\ 2 & 1 & 3 \\ 3 & 1 & 4 \\ 4 & 1 & 5 \end{pmatrix}, \qquad \mathbf{x} = \begin{pmatrix} x_1 \\ x_2 \\ x_3 \end{pmatrix}, \qquad \mathbf{b} = \begin{pmatrix} 11 \\ 3 \\ 10 \\ -1 \end{pmatrix}.$$

The matrix A is of size 4×3, and the matrix form of the system of equations is $A\mathbf{x} = \mathbf{b}$. Further, denote the columns of A by $\mathbf{a}_1, \mathbf{a}_2$, and $\mathbf{a}_3$. That is,

$$\mathbf{a}_1 = \begin{pmatrix} 1 \\ 2 \\ 3 \\ 4 \end{pmatrix}, \qquad \mathbf{a}_2 = \begin{pmatrix} 1 \\ 1 \\ 1 \\ 1 \end{pmatrix}, \qquad \mathbf{a}_3 = \begin{pmatrix} 2 \\ 3 \\ 4 \\ 5 \end{pmatrix}.$$

Now the system of equations can be rewritten as

$$x_1 \begin{pmatrix} 1 \\ 2 \\ 3 \\ 4 \end{pmatrix} + x_2 \begin{pmatrix} 1 \\ 1 \\ 1 \\ 1 \end{pmatrix} + x_3 \begin{pmatrix} 2 \\ 3 \\ 4 \\ 5 \end{pmatrix} = \begin{pmatrix} 11 \\ 3 \\ 10 \\ -1 \end{pmatrix}$$

or, equivalently, in vector form, as

$$x_1 \mathbf{a}_1 + x_2 \mathbf{a}_2 + x_3 \mathbf{a}_3 = \mathbf{b},$$

just as in Example 4.12.

∎

In general, if A is a $m \times n$ matrix with columns denoted $\mathbf{a}_1, \mathbf{a}_2, \ldots \mathbf{a}_n$ and $\mathbf{x}$ and $\mathbf{b}$ are vectors in $\mathbb{R}^m$, the system of linear equations for $\mathbf{x} = (x_1, x_2, \ldots, x_n)$ with a coefficient matrix A and right-hand side $\mathbf{b} = (b_1, b_2, \ldots, b_m)$ can be written in *matrix form* as

$$A\mathbf{x} = \mathbf{b}$$

and in *vector form* as

$$x_1 \mathbf{a}_1 + x_2 \mathbf{a}_2 + \cdots + x_n \mathbf{a}_n = \mathbf{b},$$

where the number of terms n on the left-hand side is the same as the number of variables.

It is important that you become comfortable with this notation and be able to transition easily between these forms when working with systems of linear equations. For one, they allow us to write a system of equations in a more compact way, but there is another, even more important, benefit. The vector form shows that to find a solution of the system $\mathbf{x}$, means to find weights $x_1, x_2, \ldots, x_n$ that give the vector $\mathbf{b}$ as a linear combination of the column vectors $\mathbf{a}_1, \mathbf{a}_2, \ldots \mathbf{a}_n$.

If the system does not have a solution, this would mean that $\mathbf{b}$ *cannot* be written a linear combination of the columns $\mathbf{a}_1, \mathbf{a}_2, \ldots \mathbf{a}_n$ and, thus (review Definition 4.1.2), that the vector $\mathbf{b}$ is *not* in the span of the columns $\mathbf{a}_1, \mathbf{a}_2, \ldots \mathbf{a}_n$. We will use this fact in later sections.

Exercise 4.6 Write each of the systems below in matrix form *and* in vector form, identifying clearly the matrix A and the column vectors $\{\mathbf{a}_i\}$, as in the examples above.

$$
\begin{array}{rrrrr}
& x_1 & - & 3x_2 & + & 2x_3 & = & 7 \\
1. & 2x_1 & + & x_2 & & & = & 6 \\
& 5x_1 & & & - & x_3 & = & 0
\end{array}
$$

$$
\begin{array}{rrrrrrr}
2. & x_1 & + & x_2 & - & x_3 & - & x_4 & = & -5 \\
& -x_1 & + & 3x_2 & - & 4x_3 & & & = & -5
\end{array}
$$

$$
\begin{array}{rrrrr}
& 6x_1 & - & 5x_2 & + & 2x_3 & = & 0 \\
3. & & & 2x_2 & + & x_3 & = & 4 \\
& x_1 & - & 5x_2 & + & 6x_3 & = & 12
\end{array}
$$

$$
\begin{array}{rrrrr}
& x_1 & + & 6x_2 & + & x_3 & = & 9 \\
4. & 5x_1 & - & 2x_2 & + & 8x_3 & = & 4 \\
& x_1 & + & 3x_2 & - & 2x_3 & = & 7
\end{array}
$$

Exercise 4.7 Express each matrix equation as a system of equations.

1. $\begin{pmatrix} 0 & 5 & -2 \\ 1 & 2 & 9 \\ 3 & 0 & 4 \end{pmatrix} \begin{pmatrix} x_1 \\ x_2 \\ x_3 \end{pmatrix} = \begin{pmatrix} 0 \\ 6 \\ 1 \end{pmatrix}$;

2. $\begin{pmatrix} 1 & -1 & 4 \\ 2 & 7 & 5 \end{pmatrix} \begin{pmatrix} x_1 \\ x_2 \\ x_3 \end{pmatrix} = \begin{pmatrix} 7 \\ -6 \end{pmatrix}$.

Exercise 4.8 For each of the systems in Exercise 4.6, write the system of equations in matrix form and in vector form, identifying clearly the matrix A and the vectors $\{a_i\}$, as in the examples above.

Exercise 4.9 In each case, write $Ax = b$ as a system of equations and in vector form, using the same notation we used in the examples.

1. $A = \begin{pmatrix} 1 & 0 & 2 & 1 \\ 1 & -2 & 0 & 1 \\ 3 & 1 & 5 & 0 \\ 1 & 5 & 7 & 1 \end{pmatrix}$, $\mathbf{b} = \begin{pmatrix} 0 \\ 0 \\ 1 \\ 1 \end{pmatrix}$;

2. $A = \begin{pmatrix} 0 & 1 & 3 \\ 2 & 4 & 10 \\ 3 & -2 & -9 \end{pmatrix}$, $\mathbf{b} = \begin{pmatrix} 1 \\ 5 \\ 0 \end{pmatrix}$.

4.4 Linear Independence and Bases

The concept of independence is fundamental in linear algebra. Recall from Definition 4.1.2 that a set of vectors is *linearly independent* if the only way $x_1\mathbf{a}_1 + x_2\mathbf{a}_2 + \cdots + x_n\mathbf{a}_n = \mathbf{0}$ is to have $x_1 = x_2 = \cdots = x_n = 0$.

4.4.1 Determining If a Set of Vectors is Linearly Independent

We begin with two special cases for $n = 1$ (that is, when the vector set contains a single vector) that deserve attention:

1. The set $\{\mathbf{0}\}$, containing only the zero vector, is linearly dependent since $x_1\mathbf{0} = \mathbf{0}$ for any nonzero scalar x_1.
2. A set containing a single nonzero vector is always linearly independent, as if $\mathbf{a}_1 \neq \mathbf{0}$, $x_1\mathbf{a}_1 = \mathbf{0}$ is only possible when $x_1 = 0$.

Consider now a set of vectors $\{\mathbf{a}_1, \mathbf{a}_2, \ldots, \mathbf{a}_n\}$ in $\mathbb{R}^m$. How can we check if they are linearly independent or dependent? In light of what we saw in the previous section, we can easily find the answer. We will have to:

1. Create a matrix A with the vectors $\mathbf{a}_1, \mathbf{a}_2, \ldots, \mathbf{a}_n$ as columns;
2. Check if the system of equations $A\mathbf{x} = \mathbf{0}$ has only the trivial solution $\mathbf{x} = \mathbf{0}$.
3. If $\mathbf{x} = \mathbf{0}$ is the only solution, the set is linearly independent. Otherwise, the set is linearly dependent.

■ **Example 4.14** Let $\mathbf{a}_1 = \begin{pmatrix} 1 \\ 0 \\ 2 \end{pmatrix}$, $\mathbf{a}_2 = \begin{pmatrix} 1 \\ -2 \\ 0 \end{pmatrix}$, and $\mathbf{a}_3 = \begin{pmatrix} 3 \\ -1 \\ 5 \end{pmatrix}$. Is the set $\{\mathbf{a}_1, \mathbf{a}_2, \mathbf{a}_3\}$ linearly independent or linearly dependent?

To answer this question, we will follow the three steps outlined above.

1. We first form the matrix A whose columns are $\mathbf{a}_1, \mathbf{a}_2$, and $\mathbf{a}_3$:

$$A = \begin{pmatrix} 1 & 1 & 3 \\ 0 & -2 & -1 \\ 2 & 0 & 5 \end{pmatrix}$$

2. To solve $A\mathbf{x} = \mathbf{0}$, we find the row reduced form of A (which, again, you can find using Desmos or other technology) that is

$$\begin{pmatrix} 1 & 0 & 2.5 \\ 0 & 1 & 0.5 \\ 0 & 0 & 0 \end{pmatrix}$$

This shows that x_3 is a free variable, so $A\mathbf{x} = \mathbf{0}$ has infinitely many solutions.

3. Since $\mathbf{x} = \mathbf{0}$ is *not* the only solution of $A\mathbf{x} = \mathbf{0}$, the set of vectors is linearly *dependent*.

■

■ **Example 4.15** Let $\mathbf{a}_1 = \begin{pmatrix} 2 \\ 2 \\ 1 \end{pmatrix}$, $\mathbf{a}_2 = \begin{pmatrix} -4 \\ 6 \\ 5 \end{pmatrix}$, and $\mathbf{a}_3 = \begin{pmatrix} 2 \\ 0 \\ 0 \end{pmatrix}$. Is the set $\{\mathbf{a}_1, \mathbf{a}_2, \mathbf{a}_3\}$ linearly independent or linearly dependent?

Again, we will follow the three steps outlined above to answer this question.

1. We form the matrix A whose columns are $\mathbf{a}_1, \mathbf{a}_2$, and $\mathbf{a}_3$:

$$A = \begin{pmatrix} 2 & -4 & 2 \\ 2 & 6 & 0 \\ 1 & 5 & 0 \end{pmatrix}$$

2. To solve $A\mathbf{x} = 0$, we find the row reduced form of A, which is

$$\begin{pmatrix} 1 & 0 & 0 \\ 0 & 1 & 0 \\ 0 & 0 & 1 \end{pmatrix}$$

This tells us that $\mathbf{x} = \mathbf{0}$ is the only solution of the system $A\mathbf{x} = \mathbf{0}$.

3. Since $\mathbf{x} = \mathbf{0}$ is the only solution of $A\mathbf{x} = \mathbf{0}$, the set of vectors is linearly *independent*.

■

Exercise 4.10 In each case, determine if the given vectors form a linearly independent set.

1. $\mathbf{a}_1 = \begin{pmatrix} 2 \\ 4 \\ 6 \end{pmatrix}$, $\mathbf{a}_2 = \begin{pmatrix} 3 \\ 2 \\ 9 \end{pmatrix}$, and $\mathbf{a}_3 = \begin{pmatrix} 5 \\ 2 \\ 9 \end{pmatrix}$;

2. $\mathbf{a}_1 = \begin{pmatrix} -4 \\ 6 \\ 5 \end{pmatrix}$, $\mathbf{a}_2 = \begin{pmatrix} -2 \\ 0 \\ 6 \end{pmatrix}$, and $\mathbf{a}_3 = \begin{pmatrix} -2 \\ -2 \\ -1 \end{pmatrix}$;

3. $\mathbf{a}_1 = \begin{pmatrix} -2 \\ 4 \\ 6 \\ 2 \end{pmatrix}$, $\mathbf{a}_2 = \begin{pmatrix} -1 \\ 2 \\ 4 \\ 9 \end{pmatrix}$, and $\mathbf{a}_3 = \begin{pmatrix} 5 \\ 2 \\ 1 \\ 3 \end{pmatrix}$;

4. $\mathbf{a}_1 = \begin{pmatrix} -3 \\ 4 \\ 8 \\ 6 \end{pmatrix}$, $\mathbf{a}_2 = \begin{pmatrix} -2 \\ 1 \\ 4 \\ -3 \end{pmatrix}$, $\mathbf{a}_3 = \begin{pmatrix} 5 \\ 2 \\ 1 \\ 3 \end{pmatrix}$, and $\mathbf{a}_4 = \begin{pmatrix} 1 \\ 0 \\ 1 \\ 0 \end{pmatrix}$;

5. $\mathbf{a}_1 = \begin{pmatrix} 5 \\ 2 \end{pmatrix}$, $\mathbf{a}_2 = \begin{pmatrix} -1 \\ 3 \end{pmatrix}$, and $\mathbf{a}_3 = \begin{pmatrix} -7 \\ 2 \end{pmatrix}$.

Exercise 4.11 Think about the following: If you were to pick at random three vectors in $\mathbb{R}^3$, do you think you would be more likely to pick a linearly dependent or a linearly independent set? (In your standard linear algebra class, you will learn how to answer this question in a rigorous way.) ▪

4.4.2 What Is a Basis?

Recall from Definition 4.1.2 that a *basis* for a set of vectors V is a collection of vectors that are linearly independent <u>and</u> span V.

To say this in a different way, a basis provides the "building blocks" of a vector space. It is like building blocks of a constriction toy such as Legos or tinker toys. The manufacturer must include all the necessary shapes to build the toys in the manual – cubes, spheres, circles, figurines, etc. If the shapes provided are not enough, customers will be dissatisfied. If there are extra shapes that can be built from simpler ones, the manufacturer's cost will be higher. With this analogy, the vectors in a basis will be the shapes included in the construction set. Every toy in the manual can be viewed as a linear combination of shapes – e.g., to build a truck with a truck driver, you may need fifteen 1" cubes, 4 circles, and 1 driver figurine. Spanning gives enough shapes to build the toys. Linear independence ensures that no shape in the set can be built from the remaining shapes (e.g., a 2" cube would be unnecessary, as it can be built from eight 1" cubes).

A basis provides the most efficient way for describing a vectors space. Every vector in a space V is a linear combination of the basis vectors, and the number of vectors in the basis is as small as possible. That is to say, removing any one vector from the basis would mean that the remaining vectors no longer span V.

The most commonly used basis of m-dimensional vectors in $\mathbb{R}^m$ is the set of vectors

$$\mathbf{e}_1 = \begin{pmatrix} 1 \\ 0 \\ 0 \\ \vdots \\ 0 \end{pmatrix}, \quad \mathbf{e}_2 = \begin{pmatrix} 0 \\ 1 \\ 0 \\ \vdots \\ 0 \end{pmatrix}, \quad \mathbf{e}_3 = \begin{pmatrix} 0 \\ 0 \\ 1 \\ \vdots \\ 0 \end{pmatrix}, \ldots, \quad \mathbf{e}_m = \begin{pmatrix} 0 \\ 0 \\ 0 \\ \vdots \\ 1 \end{pmatrix}.$$

It is easy to see that the vectors are linearly independent: If A is the $m \times m$ matrix A with columns $\mathbf{e}_1, \mathbf{e}_2, \ldots, \mathbf{e}_m$, the system of equations $A\mathbf{x} = \mathbf{0}$ has only the trivial solution $\mathbf{x} = \mathbf{0}$. We can see that they also span $\mathbb{R}^m$ because any vector $\mathbf{b} = (b_1, b_2, \ldots, b_m)$ can be written as a linear combination

$$\mathbf{b} = b_1\mathbf{e}_1 + b_2\mathbf{e}_2 + \cdots + b_m\mathbf{e}_m.$$

In your conventional linear algebra course you will see that every basis of a particular vector space has the same number of vectors. That number of vectors provides the formal definition for the *dimension* of a vector space.

Definition 4.4.1 The number of vectors in any basis of a vector space V defines the *dimension* of V.

5 Column Space and Null Space of a matrix A

When we introduced the vector form for a system of linear equations $A\mathbf{x} = \mathbf{b}$, we made the following important observation: to find all solutions of this system is the same as to find all ways in which $\mathbf{b}$ can be written as a linear combination

$$\mathbf{b} = x_1\mathbf{a}_1 + x_2\mathbf{a}_2 + \ldots x_n\mathbf{a}_n$$

where $\mathbf{a}_1, \mathbf{a}_2, \ldots \mathbf{a}_n$ are the columns of the matrix A. A unique solution would mean that there is a unique way to write $\mathbf{b}$ as a linear combination of $\mathbf{a}_1, \mathbf{a}_2, \ldots \mathbf{a}_n$. When there are infinitely many solutions, there will be infinitely many ways to do so. When the system does not have a solution, this shows that such representation is not possible, so **b** is not in the span of $\mathbf{a}_1, \mathbf{a}_2, \ldots \mathbf{a}_n$. Due to the central role the span of $\{\mathbf{a}_1, \mathbf{a}_2, \ldots, \mathbf{a}_n\}$ plays, it is given a special name.

4.5.1 Defining Column Space and the Null Space

> **Definition 4.5.1** Let A be a matrix with columns $\mathbf{a}_1, \mathbf{a}_2, \ldots \mathbf{a}_n$. The span of $\{\mathbf{a}_1, \mathbf{a}_2, \ldots, \mathbf{a}_n\}$ forms the *column space of* A and is denoted by $C(A)$.

In your standard linear algebra class you will show that the set $C(A)$ is indeed a vector space (that is, it satisfies all axioms in Definition 4.1.2), which justifies its name.

■ **Example 4.16** Consider the coefficient matrix A from Example 4.12 and the set of all linear combinations $x_1\mathbf{a}_1 + x_2\mathbf{a}_2 + x_3\mathbf{a}_3$ of its columns $\mathbf{a}_1, \mathbf{a}_2$, and $\mathbf{a}_3$. The resulting set of vectors in $\mathbb{R}^3$ gives the column space $C(A)$ of the matrix A.   ■

In the special case when $\mathbf{b} = \mathbf{0}$, the system $A\mathbf{x} = \mathbf{0}$ always has the solution $\mathbf{x} = \mathbf{0}$, the so-called *trivial solution*. Thus, there are only two possibilities for $A\mathbf{x} = \mathbf{0}$: either $x = \mathbf{0}$ is the only solution or there are infinitely many others.

> **Definition 4.5.2** The set of solutions for a systems of linear equations $A\mathbf{x} = \mathbf{0}$ is called the *null space* of the matrix A and is denoted by $N(A)$.

With this terminology, the following statements are equivalent:
- The only solution of $A\mathbf{x} = \mathbf{0}$ is $\mathbf{x} = \mathbf{0}$;
- The null space $N(A)$ contains only the zero vector $\mathbf{0}$;

A third, equivalent way to say the same thing is to use linear independence: When $\mathbf{x} = \mathbf{0}$ is the only solution of $A\mathbf{x} = \mathbf{0}$, the system of equations (in vector form)

$$x_1\mathbf{a}_1 + x_2\mathbf{a}_2 + \ldots x_n\mathbf{a}_n = \mathbf{0} \quad \text{has only the solution} \quad x_1 = 0, x_2 = 0, \ldots, x_n = 0.$$

From Definition 4.1.2, we know that this means that the columns of A form a linearly independent set.

Thus, the following three statements are equivalent:
- The only solution of $A\mathbf{x} = \mathbf{0}$ is $\mathbf{x} = \mathbf{0}$;
- The null space $N(A)$ contains only the zero vector $\mathbf{0}$;
- The columns of A form a linearly independent set $\{\mathbf{a}_1, \mathbf{a}_2, \ldots, \mathbf{a}_n\}$.

4.5.2 A Closer Look at $C(A)$ and $N(A)$ and Their Bases

Given a set of vectors, we already know how to determine if it is linearly independent. In this section we consider several important problems that will allow us to determine bases for the columns space $C(A)$ and the null space $N(A)$ of a matrix A.

Let A be a $m \times n$ matrix, $\mathbf{x}$ be a vector in $\mathbb{R}^n$, and $\mathbf{b}$ be a vector in $\mathbb{R}^m$. Denote the columns of A by $\mathbf{a}_1, \mathbf{a}_2, \ldots, \mathbf{a}_n$, as before. Consider the system of equations $A\mathbf{x} = \mathbf{b}$.

We consider the following problems:

Problem 1. We want to know the set of vectors $\mathbf{b}$, for which the system $A\mathbf{x} = \mathbf{b}$ has a solution.

Problem 2. We want to know how to find a basis for $C(A)$.

Problem 3. We want to find the null space $N(A)$ and a basis for $N(A)$.

We now use several examples to illustrate how we can solve these problems. The reduced echelon form of A helps find the answers.

■ **Example 4.17** Let's consider the matrix

$$A = \begin{pmatrix} 1 & 1 & 2 \\ 2 & 1 & 3 \\ 3 & 1 & 4 \\ 4 & 1 & 5 \end{pmatrix}.$$

Here $m = 4$ and $n = 3$.

Problem 1 Solution: We want to find all vectors $\mathbf{b} = (b_1, b_2, b_3, b_4)$, for which there is a solution $\mathbf{x} = \begin{pmatrix} x_1 \\ x_2 \\ x_3 \end{pmatrix}$ for the system

$$A\mathbf{x} = \begin{pmatrix} 1 & 1 & 2 \\ 2 & 1 & 3 \\ 3 & 1 & 4 \\ 4 & 1 & 5 \end{pmatrix} \begin{pmatrix} x_1 \\ x_2 \\ x_3 \end{pmatrix} = \begin{pmatrix} b_1 \\ b_2 \\ b_3 \\ b_4 \end{pmatrix} = \mathbf{b}.$$

We first obtain the reduced row echelon form of A, which we will call U:

$$\begin{pmatrix} 1 & 1 & 2 \\ 2 & 1 & 3 \\ 3 & 1 & 4 \\ 4 & 1 & 5 \end{pmatrix} \rightarrow \begin{pmatrix} 1 & 0 & 1 \\ 0 & 1 & 1 \\ 0 & 0 & 0 \\ 0 & 0 & 0 \end{pmatrix} = U.$$

We see that x_1 and x_2 are leading variables and x_3 is a free variable. For the system to be consistent, there must be a vector $\mathbf{x} = (x_1, x_2, x_3)$ for which $\mathbf{b}$ can be written as $x_1\mathbf{a}_1 + x_2\mathbf{a}_2 + x_3\mathbf{a}_3 = \mathbf{b}$. Since x_3 is free, we can set $x_3 = 0$. So, $\mathbf{x} = (x_1, x_2, 0)$ is a solution. Thus, the set of all vectors $\mathbf{b}$ for which $A\mathbf{x} = \mathbf{b}$ is the set of all linear combinations $\mathbf{b} = x_1\mathbf{a}_1 + x_2\mathbf{a}_2$ of the columns $\mathbf{a}_1$ and $\mathbf{a}_2$ (the columns with leading 1's).

Problem 2 solution: Will, in general, the set of all linear combinations $\mathbf{b} = x_1\mathbf{a}_1 + x_2\mathbf{a}_2 + x_3\mathbf{a}_3$ for values of $x_3 \neq 0$ give us more vectors $\mathbf{b}$ than the span of the columns with leading 1's? It won't because each column that corresponds to a free variable can be written as a linear combination of the columns with leading 1's! This means that the set $\{\mathbf{a}_1, \mathbf{a}_2\}$ spans $C(A)$.

We can also see easily that the columns $\mathbf{a}_1$ and $\mathbf{a}_2$ form a linearly independent set: We consider a matrix B with columns $\mathbf{a}_1$ and $\mathbf{a}_2$ and check if its null space contains only the vector $\mathbf{x} = \mathbf{0}$. We row-reduce the matrix B and obtain

$$B = \begin{pmatrix} 1 & 1 \\ 2 & 1 \\ 3 & 1 \\ 4 & 1 \end{pmatrix} \rightarrow \begin{pmatrix} 1 & 0 \\ 0 & 1 \\ 0 & 0 \\ 0 & 0 \end{pmatrix}.$$

Thus $B\mathbf{x} = \mathbf{0}$ has only the trivial solution and the columns $\mathbf{a}_1$ and $\mathbf{a}_2$ are linearly independent.

Because $\{\mathbf{a}_1, \mathbf{a}_2\}$ is a linearly independent set and spans the columns space $C(A)$, $\{\mathbf{a}_1, \mathbf{a}_2\}$ is a basis for $C(A)$.

Problem 3 Solution: We again use the row reduced matrix U of A. Since x_3 is free, the solution of the system $A\mathbf{x} = \mathbf{0}$ is $x_1 = -x_3$ and $x_2 = -x_3$. As a free variable, x_3 can take any value, so let $x_3 = t$, where t is a real number. Then all solutions of the system $A\mathbf{x} = \mathbf{0}$ are of the form

$$\mathbf{x} = \begin{pmatrix} x_1 \\ x_2 \\ x_3 \end{pmatrix} = \begin{pmatrix} -t \\ -t \\ t \end{pmatrix} = t \begin{pmatrix} -1 \\ -1 \\ 1 \end{pmatrix}.$$

Since t can be any number, the null space of the matrix A contains all scalar multiples of the vector $(-1, -1, 1)$.

To check our work, we note that

$$\begin{pmatrix} 1 & 1 & 2 \\ 2 & 1 & 3 \\ 3 & 1 & 4 \\ 4 & 1 & 5 \end{pmatrix} \begin{pmatrix} -t \\ -t \\ t \end{pmatrix} = \begin{pmatrix} -t - t + 2t \\ -2t - t + 3t \\ -3t - t + 4t \\ -4t - t + 5t \end{pmatrix} = \begin{pmatrix} 0 \\ 0 \\ 0 \\ 0 \end{pmatrix}.$$

So, the scalar multiples of $\begin{pmatrix} -1 \\ -1 \\ 1 \end{pmatrix}$ span the null space $N(A)$ of A. Further,

since we know from Section 4.4.1 that a single nonzero vector always forms a

linearly independent set, we see that $\begin{pmatrix} -1 \\ -1 \\ 1 \end{pmatrix}$ is a basis for $N(A)$.   ■

■ **Example 4.18** Let

$$A = \begin{pmatrix} 1 & 2 & 1 & 1 & 7 \\ 1 & 2 & 2 & -1 & 12 \\ 2 & 4 & 0 & 6 & 4 \end{pmatrix}.$$

Check that its reduced echelon form U is

$$U = \begin{pmatrix} 1 & 2 & 0 & 3 & 2 \\ 0 & 0 & 1 & 2 & 5 \\ 0 & 0 & 0 & 0 & 0 \end{pmatrix}.$$

We can see that columns $\mathbf{a}_1$ and $\mathbf{a}_3$ are columns with leading 1's and x_2, x_4, and x_5 are free variables.

Solution to Problem 1: Write the system in vector form:

$$x_1\mathbf{a}_1 + x_2\mathbf{a}_2 + x_3\mathbf{a}_3 + x_4\mathbf{a}_4 + x_5\mathbf{a}_5 = \mathbf{b}.$$

We can take any values for the free variables x_2, x_4, x_5 and the simplest case would be to set them equal to zero. This shows that when $\mathbf{b}$ is in the span of $\{\mathbf{a}_1, \mathbf{a}_3\}$, the system will have a solution.

Problem 2 Solution: As in the previous example, we can check that the set $\{\mathbf{a}_1, \mathbf{a}_3\}$ is linearly independent, and that each column corresponding to a free variable can be written as a linear combination of $\mathbf{a}_1$ and $\mathbf{a}_3$ (verify!). So, just as in our previous example, we see that $\{\mathbf{a}_1, \mathbf{a}_3\}$ is a basis for $C(A)$.

Problem 3 Solution: Since x_2, x_4, and x_5 are free variables, $A\mathbf{x} = \mathbf{0}$ will have infinitely many solutions. If s, r, t are arbitrary numbers, we can set $x_2 = r$, $x_4 = s$, and $x_5 = t$, then solve for x_1 and x_3: $x_1 = -2x_2 - 3x_4 - 2x_5 = -2r - 3s - 2t$, $x_3 = -2x_4 - 5x_5 = -2s - 5t$. This gives the null space of A; that is, the solution set for $A\mathbf{x} = \mathbf{0}$.

To find a basis for $N(A)$, we then can write the solution set as follows:

$$\mathbf{x} = \begin{pmatrix} x_1 \\ x_2 \\ x_3 \\ x_4 \\ x_5 \end{pmatrix} = \begin{pmatrix} -2r - 3s - 2t \\ r \\ -2s - 5t \\ s \\ t \end{pmatrix} = \begin{pmatrix} -2r & - & 3s & - & 2t \\ r & + & 0s & + & 0t \\ 0r & - & 2s & - & 5t \\ 0r & + & s & + & 0t \\ 0r & + & 0s & + & t \end{pmatrix}$$

$$= r \begin{pmatrix} -2 \\ 1 \\ 0 \\ 0 \\ 0 \end{pmatrix} + s \begin{pmatrix} -3 \\ 0 \\ -2 \\ 1 \\ 0 \end{pmatrix} + t \begin{pmatrix} -2 \\ 0 \\ -5 \\ 0 \\ 1 \end{pmatrix},$$

where r, s, and t are arbitrary numbers. This tells us that the solution set of the system $A\mathbf{x} = \mathbf{0}$ is the set of all linear combinations of the vectors $\mathbf{u}$, $\mathbf{v}$, $\mathbf{w}$ where

$$\mathbf{u} = \begin{pmatrix} -2 \\ 1 \\ 0 \\ 0 \\ 0 \end{pmatrix}, \quad \mathbf{v} = \begin{pmatrix} -3 \\ 0 \\ -2 \\ 1 \\ 0 \end{pmatrix}, \quad \mathbf{w} = \begin{pmatrix} -2 \\ 0 \\ -5 \\ 0 \\ 1 \end{pmatrix}.$$

Put another way, the vectors $\mathbf{u}, \mathbf{v}$, and $\mathbf{w}$ span $N(A)$.

The vectors $\mathbf{u}, \mathbf{v}$, and $\mathbf{w}$ will form a basis if they form a linearly independent set. Verify this on your own by row-reducing the matrix with columns $\mathbf{u}, \mathbf{v}$, and $\mathbf{w}$ to prove that $\{\mathbf{u}, \mathbf{v}, \mathbf{w}\}$ is a basis for $N(A)$.

∎

■ **Example 4.19** Consider a matrix A whose row reduced form is

$$U = \begin{pmatrix} 1 & 0 & 4 & 1 \\ 0 & 1 & 2 & 2 \\ 0 & 0 & 0 & 0 \\ 0 & 0 & 0 & 0 \end{pmatrix}.$$

Just as in the previous examples, this tells us that the columns with leading 1's – $\mathbf{a}_1$ and $\mathbf{a}_2$ form a linearly independent set of columns that span $C(A)$. Thus $\{\mathbf{a}_1, \mathbf{a}_2\}$ is a basis for $C(A)$.

To find the null space $N(A)$, we need to solve

$$\begin{pmatrix} 1 & 0 & 4 & 1 \\ 0 & 1 & 2 & 2 \\ 0 & 0 & 0 & 0 \\ 0 & 0 & 0 & 0 \end{pmatrix} \begin{pmatrix} x_1 \\ x_2 \\ x_3 \\ x_4 \end{pmatrix} = \begin{pmatrix} 0 \\ 0 \\ 0 \\ 0 \end{pmatrix}.$$

Now x_3 and x_4 are free variables and they can take any values $x_3 = s$, $x_4 = t$. We can express x_1 and x_2 in terms of x_3 and x_4:

$$\begin{array}{rl}
x_1 + 4s + t & = 0 \\
x_2 + 2s + 2t & = 3
\end{array} \quad \text{or, equivalently,} \quad \begin{array}{rl}
x_1 = & -4s - t \\
x_2 = & -2s - 2t
\end{array}$$

We can now write:

$$\mathbf{x} = \begin{pmatrix} x_1 \\ x_2 \\ x_3 \\ x_4 \end{pmatrix} = \begin{pmatrix} -4s - t \\ -2s - 2t \\ s \\ t \end{pmatrix} = \begin{pmatrix} -4s & - & t \\ -2s & - & 2t \\ s & + & 0t \\ 0s & + & t \end{pmatrix} = s \begin{pmatrix} -4 \\ -2 \\ 1 \\ 0 \end{pmatrix} + t \begin{pmatrix} -1 \\ -2 \\ 0 \\ 1 \end{pmatrix},$$

where s and t can be any real numbers. This shows that the null space $N(A)$ is the span of the vectors

$$\mathbf{u} = \begin{pmatrix} -4 \\ -2 \\ 1 \\ 0 \end{pmatrix} \quad \text{and} \quad \mathbf{v} = \begin{pmatrix} -1 \\ -2 \\ 0 \\ 1 \end{pmatrix}.$$

Just as in the previous examples, the set of vectors $\{\mathbf{u}, \mathbf{v}\}$ is linearly independent (verify!). Thus, $\{\mathbf{u}, \mathbf{v}\}$ is a basis for $N(A)$. ∎

The last several examples illustrate a few results that are fundamental in linear algebra. You will see their proofs in the general case in your conventional linear algebra course. Here, we simply note that what we saw in the examples above generalizes readily for the arbitrary case: If A is a $m \times n$ matrix, and $\mathbf{b}$ is in $\mathbb{R}^m$, the solutions to the three fundamental problems we stated above can be found by following these steps:

1. Find the row reduced echelon form U of the matrix A.
2. To solve **Problem 1**, consider the columns of A corresponding to leading 1's in U. This subset of columns spans the column space $C(A)$. The span of these columns is also the set of all vectors $\mathbf{b}$ for which the system $A\mathbf{x} = \mathbf{b}$ has a solution.
3. The set of columns of A corresponding to leading 1's in U gives a solution to **Problem 2**, too. It is a *linearly independent set* that spans $C(A)$, so it is a basis for $C(A)$.

4. To solve **Problem 3**, we again use the row reduced echelon form U to solve the system $A\mathbf{x} = \mathbf{0}$. If A has k columns with leading 1's, the remaining $n - k$ columns correspond to free variables. The null space $N(A)$ is spanned by exactly $n - k$ vectors, which are determined from the general solution of the system $A\mathbf{x} = \mathbf{0}$. These $n - k$ vectors are linearly independent and, thus, provide a basis for the null space $N(A)$.

A couple of more important observations follow from here:

5. The dimension of $C(A)$ (the number of vectors in any basis for $C(A)$) is always equal to the number of columns with leading 1's.
6. The dimension of $N(A)$ (the number of vectors in any basis for $N(A)$) is always equal to the number of columns corresponding to free variables.
7. The dimension of the column space $C(A)$ and the dimension of the null space $N(A)$ always add up to the number of columns n of the matrix A.

Using what we have learned from solving Problems 1, 2, and 3 above, we can now solve two other important problems:

Problem 4: When is $C(A) = \mathbb{R}^m$? Of, equivalently, when does $A\mathbf{x} = \mathbf{b}$ have a solution for *all* $\mathbf{b}$ in $\mathbb{R}^m$?

Problem 4 solution: To have the column space be the set of all vectors in $\mathbb{R}^m$, $C(A)$ must have a basis of m vectors. Thus, the matrix A must have *at least m* columns corresponding to leading 1's.

Problem 5: When does $A\mathbf{x} = \mathbf{b}$ have a *unique solution* for *all* $\mathbf{b}$ in $\mathbb{R}^m$?

Problem 5 solution: If we want to have a unique solution for every $\mathbf{b}$ in $\mathbb{R}^m$, the matrix A must have m columns corresponding to leading 1's and no free variables. This means A must be a square matrix with $n = m$ columns and have a row reduced form that is the $m \times m$ identity matrix.

■ **Example 4.20** Consider the matrix A and its row reduced form:

$$A = \begin{pmatrix} -1 & 2 & -1 & 1 \\ 1 & 3 & 0 & 1 \\ 1 & 5 & 0 & 1 \end{pmatrix} \rightarrow \begin{pmatrix} 1 & 0 & 0 & 1 \\ 0 & 1 & 0 & 0 \\ 0 & 0 & 1 & -2 \end{pmatrix}.$$

Here $m = 3$ and $n = 4$. We have 3 columns with leading 1's and one free variable. The first three columns of A form a basis for $C(A)$ and $C(A) = \mathbb{R}^3$. The system has a solution for any $\mathbf{b}$ in $\mathbb{R}^3$. However, since we have a free variable, $A\mathbf{x} = \mathbf{b}$ has infinitely many solutions.   ■

■ **Example 4.21** Consider the following matrix and its row reduced form:

$$A = \begin{pmatrix} 2 & -4 & 2 \\ 2 & 6 & 0 \\ 1 & 5 & 0 \end{pmatrix} \rightarrow \begin{pmatrix} 1 & 0 & 0 \\ 0 & 1 & 0 \\ 0 & 0 & 1 \end{pmatrix}.$$

In this case, $m = n = 3$, and the row reduced matrix of A is the identity matrix. This means that $Ax = b$ has a *unique solution* for any b in $\mathbb{R}^3$. ■

■ **Example 4.22** Consider the matrices whose row reduced forms are given below. In each case, determine if $C(A) = \mathbb{R}^m$ and explain. In each case, determine also the dimension of the null space $N(A)$.

1. $\begin{pmatrix} 1 & 2 & 0 \\ 0 & 0 & 1 \\ 0 & 0 & 0 \\ 0 & 0 & 0 \end{pmatrix}.$

In this case, $m = 4$, and we have only two columns with leading 1's – too few to span $\mathbb{R}^m$. Thus $C(A) \neq \mathbb{R}^4$. There is one free variable, so the dimension of $N(A)$ is 1.

2. $\begin{pmatrix} 1 & 0 & 0 & 0 & 0 & 2 \\ 0 & 1 & 1 & 0 & 0 & 1 \\ 0 & 0 & 0 & 1 & 0 & 2 \\ 0 & 0 & 0 & 0 & 1 & 4 \end{pmatrix}.$

In this case, $m = 4$, and we have four columns with leading 1's. Thus, $C(A) = \mathbb{R}^4$. Further, there are two free variables, so the dimension of $N(A)$ is 2.

3. $\begin{pmatrix} 1 & 0 & 0 & 0 \\ 0 & 1 & 0 & 0 \\ 0 & 0 & 1 & 0 \\ 0 & 0 & 0 & 0 \end{pmatrix}.$ In this case, $m = 4$, and there are only three columns

corresponding to leading 1's. Thus, $C(A) \neq \mathbb{R}^4$. There is one free variable, so the dimension of $N(A)$ is 1.

4. $\begin{pmatrix} 1 & 0 & 0 & 0 \\ 0 & 1 & 0 & 0 \\ 0 & 0 & 1 & 0 \\ 0 & 0 & 0 & 1 \end{pmatrix}$. In this case, $m = 4$, and there are four columns with

leading 1's. Thus, $C(A) = \mathbb{R}^4$. There are no free variables, so $A\mathbf{x} = \mathbf{0}$ has only the trivial solution. The dimension of $N(A)$ is zero. ∎

Exercise 4.12 Let

$$A = \begin{pmatrix} 2 & -2 & 4 & -2 \\ 2 & 1 & 10 & 7 \\ -4 & 4 & -8 & 4 \\ 4 & -1 & 14 & 6 \end{pmatrix}.$$

Denote its columns by $\mathbf{a}_1$, $\mathbf{a}_2$, $\mathbf{a}_3$, and $\mathbf{a}_4$.
  1. Check that the row reduced form for the matrix A is

  $$U = \begin{pmatrix} 1 & 0 & 4 & 0 \\ 0 & 1 & 2 & 0 \\ 0 & 0 & 0 & 1 \\ 0 & 0 & 0 & 0 \end{pmatrix}.$$

  2. Why does the matrix U show that the set of all vectors $\mathbf{b}$ for which $A\mathbf{x} = \mathbf{b}$ is consistent is the span of $\mathbf{a}_1$, $\mathbf{a}_2$, and $\mathbf{a}_4$.
  3. Why does the matrix U show that $A\mathbf{x} = \mathbf{0}$ has infinitely many solutions?
  4. Show that the null space of A is the set of all scalar multiples of the vector

  $$\mathbf{u} = \begin{pmatrix} -4 \\ -2 \\ 0 \\ 1 \end{pmatrix}.$$

Exercise 4.13 Let

$$A = \begin{pmatrix} 1 & 0 & 2 & 0 \\ 3 & 5 & 6 & 15 \\ -2 & 1 & -4 & 3 \end{pmatrix}.$$

Denote its columns by a_1, a_2, a_3, and a_4.

1. Check that the row reduced form for the matrix A is

$$U = \begin{pmatrix} 1 & 0 & 2 & 0 \\ 0 & 1 & 0 & 3 \\ 0 & 0 & 0 & 0 \end{pmatrix}.$$

2. Why does this tell us that $\begin{pmatrix} 1 \\ 3 \\ -2 \end{pmatrix}$ and $\begin{pmatrix} 0 \\ 5 \\ 1 \end{pmatrix}$ span C(A)?

3. Why does the row reduced form U tell us that $A\mathbf{x} = \mathbf{0}$ has infinitely many solutions?

4. Show that the vectors $\mathbf{u} = \begin{pmatrix} -2 \\ 0 \\ 1 \\ 0 \end{pmatrix}$ and $\mathbf{v} = \begin{pmatrix} 0 \\ -3 \\ 0 \\ 1 \end{pmatrix}$ span the null

space N(A). ∎

Exercise 4.14 Let A be as in Exercise 4.13 and $\mathbf{b} = \begin{pmatrix} 1 \\ 3 \\ 5 \end{pmatrix}$. Verify the

following:

1. The system of equations $A\mathbf{x} = \mathbf{b}$ does not have a solution.

2. The vector $\mathbf{b}$ is *not* in the span of the columns $a_1 = \begin{pmatrix} 1 \\ 3 \\ -2 \end{pmatrix}$ and

$a_2 = \begin{pmatrix} 0 \\ 5 \\ 1 \end{pmatrix}$. To do this, try to find numbers x_1 and x_2 such that

$x_1\mathbf{a}_1 + x_2\mathbf{a}_2 = \mathbf{b}$. That is, try to solve the system

$$x_1 \begin{pmatrix} 1 \\ 3 \\ -2 \end{pmatrix} + x_2 \begin{pmatrix} 0 \\ 5 \\ 1 \end{pmatrix} = \begin{pmatrix} 1 \\ 3 \\ 5 \end{pmatrix}$$

and show that it doesn't have a solution.

Exercise 4.15 For each matrix A, find:
- the set of all vectors $\mathbf{b}$ for which the system $A\mathbf{x} = \mathbf{b}$ has a solution;
- a basis for the column space $C(A)$;
- a basis for the null space $N(A)$;
- the dimension of $C(A)$;
- the dimension of $N(A)$.

1. $A = \begin{pmatrix} 1 & 0 & 2 & 1 \\ 1 & -2 & 0 & 1 \\ 3 & 1 & 5 & 0 \\ 1 & 5 & 7 & 1 \end{pmatrix}$

2. $A = \begin{pmatrix} 0 & 1 & 3 \\ 2 & 4 & 10 \\ 3 & -2 & -9 \end{pmatrix}$

4.5.3 Determining if a Set of Vectors Is a Basis for $\mathbb{R}^m$

In providing an answer to Problem 5, we have now arrived at a very important result: To say that the system $A\mathbf{x} = \mathbf{b}$ has a unique solution for every $\mathbf{b}$ in $\mathbb{R}^m$ is the same as to say that the columns of A form a basis for $\mathbb{R}^m$.

Several results follow from here for a $m \times n$ matrix A:

1. If A has more rows than columns (that is, if $m > n$), the columns of A cannot form a basis for $\mathbb{R}^m$. There are not enough columns with leading 1's and they can't span $\mathbb{R}^m$.
2. If a matrix A has more columns than rows (that is, if $n > m$), the columns of A cannot form a basis for $\mathbb{R}^m$. There are too many columns, so there are free variable, and the columns are not linearly independent.
3. If A is a square matrix with m rows and columns, we need to check if the columns are linearly independent. If so, the columns of A form a basis for $\mathbb{R}^m$

Thus, to check if any set of vectors $\{a_1, a_2, \ldots, a_n\}$ is a basis for $\mathbb{R}^m$, we need to follow these steps:

1. If $m \neq n$, the set is not a basis.
2. If $m = n$, create the $m \times m$ matrix A with columns $a_1, a_2, \ldots, a_m$. Row reduce the matrix A.
   - If all columns of A correspond to leading 1's, the set $\{a_1, a_2, \ldots, a_m\}$ is basis for $\mathbb{R}^m$.
   - If there are free variables, the set $\{a_1, a_2, \ldots, a_m\}$ is not a basis for $\mathbb{R}^m$.

■ **Example 4.23** In each case, determine if the set of vectors forms a basis for the indicated $\mathbb{R}^m$.

1. Consider a_1, a_2 for $\mathbb{R}^3$ with

$$a_1 = \begin{pmatrix} 4 \\ 3 \\ 6 \end{pmatrix} \quad \text{and} \quad a_2 = \begin{pmatrix} -6 \\ -2 \\ 0 \end{pmatrix}.$$

Here $m = 3$, and we only have $n = 2$ vectors, so too few for a basis.

2. Consider a_1, a_2, a_3, a_4 for $\mathbb{R}^3$ with

$$a_1 = \begin{pmatrix} 4 \\ 3 \\ 6 \end{pmatrix}, \quad a_2 = \begin{pmatrix} -6 \\ -2 \\ 0 \end{pmatrix}, \quad a_3 = \begin{pmatrix} 1 \\ 9 \\ 80 \end{pmatrix}, \quad \text{and} \quad a_4 = \begin{pmatrix} 87 \\ 55 \\ 12 \end{pmatrix}.$$

Here again $m = 3$, but we have $n = 4$ vectors, so too many for a basis.

3. Consider a_1, a_2, a_3 for $\mathbb{R}^3$ with

$$a_1 = \begin{pmatrix} 1 \\ 1 \\ 2 \end{pmatrix}, \quad a_2 = \begin{pmatrix} 3 \\ 1 \\ 0 \end{pmatrix}, \quad \text{and} \quad a_3 = \begin{pmatrix} -1 \\ 3 \\ -1 \end{pmatrix}.$$

Here $m = n$, so we form the matrix A with the vectors as columns and row-reduce:

$$A = \begin{pmatrix} 1 & 3 & -1 \\ 1 & 1 & 3 \\ 2 & 0 & -1 \end{pmatrix} \rightarrow \begin{pmatrix} 1 & 0 & 0 \\ 0 & 1 & 0 \\ 0 & 0 & 1 \end{pmatrix}.$$

All columns correspond to leading 1's, so a_1, a_2, a_3 is a basis.

4. Consider $\mathbf{a}_1, \mathbf{a}_2, \mathbf{a}_3$ for $\mathbb{R}^3$ with

$$\mathbf{a}_1 = \begin{pmatrix} 4 \\ 3 \\ 6 \end{pmatrix}, \quad \mathbf{a}_2 = \begin{pmatrix} 6 \\ -2 \\ 0 \end{pmatrix}, \quad \text{and} \quad \mathbf{a}_3 = \begin{pmatrix} 10 \\ 1 \\ 6 \end{pmatrix}.$$

Here again $m = n$, so we form the matrix A with the vectors as columns and row-reduce:

$$A = \begin{pmatrix} 4 & 6 & 10 \\ 3 & -2 & 1 \\ 6 & 0 & 6 \end{pmatrix} \rightarrow \begin{pmatrix} 1 & 0 & \frac{6}{13} \\ 0 & 1 & \frac{18}{13} \\ 0 & 0 & 0 \end{pmatrix}.$$

Since there are only two columns with leading 1's, there is a free variable, and $\{\mathbf{a}_1, \mathbf{a}_2, \mathbf{a}_3\}$ is not a basis for $\mathbb{R}^3$.

■

Exercise 4.16 For the following sets of vectors, determine whether the set forms a basis for $\mathbb{R}^3$. If it does not, explain why.

1. $\begin{pmatrix} 3 \\ 1 \\ 0 \end{pmatrix}, \begin{pmatrix} 2 \\ 5 \\ 1 \end{pmatrix}, \begin{pmatrix} 7 \\ 11 \\ 3 \end{pmatrix}, \begin{pmatrix} 1 \\ -4 \\ 2 \end{pmatrix}$;

2. $\begin{pmatrix} 1 \\ 0 \\ -1 \end{pmatrix}, \begin{pmatrix} 5 \\ 1 \\ 2 \end{pmatrix}$;

3. $\begin{pmatrix} 4 \\ 2 \\ 5 \end{pmatrix}, \begin{pmatrix} 3 \\ 1 \\ 1 \end{pmatrix}, \begin{pmatrix} 11 \\ -2 \\ 9 \end{pmatrix}$;

4. $\begin{pmatrix} 7 \\ 0 \\ 0 \end{pmatrix}, \begin{pmatrix} 2 \\ 1 \\ 5 \end{pmatrix}, \begin{pmatrix} 0 \\ -1 \\ -2 \end{pmatrix}$;

5. $\begin{pmatrix} 1 \\ 0 \\ 0 \end{pmatrix}, \begin{pmatrix} 1 \\ 2 \\ 0 \end{pmatrix}, \begin{pmatrix} 1 \\ 2 \\ 3 \end{pmatrix}$.

.4 Linear Transformations

The meaning of the formal definition of a linear transformation $T : \mathbb{R}^n \to \mathbb{R}^m$ in Definition 4.1.2 may not be immediately clear. But if we think of the function T as "transforming" (moving) each input vector $\mathbf{x}$ to an output vector $\mathbf{y}$, the following fundamental result in linear algebra tells us that we obtain $\mathbf{y}$ from multiplying $\mathbf{x}$ by a matrix:

> **Theorem 4.5.1** Given a linear transformation T from $\mathbb{R}^n$ to $\mathbb{R}^m$, there is a unique $m \times n$ matrix A such that
>
> $$y = T(\mathbf{x}) = A\mathbf{x}, \qquad \text{for any } \mathbf{x} \text{ in } \mathbb{R}^n.$$

The importance of this result would be difficult to overstate! It says that *every* linear transformation is, in fact, a multiplication by a matrix A that is unique to the linear transformation.

The details of proving this result are not so important at this point. It is more important to realize that it allows us to answer some important questions about linear transformations by linking them to questions about solving systems of linear equations, which we have already answered.

Section 6.8 of the Appendix provides the relevant terminology and definitions regarding functions, which you may want to review. Here, we have a function $T : \mathbb{R}^n \to \mathbb{R}^m$, for which $\mathbb{R}^n$ is the domain and $\mathbb{R}^m$ is the co-domain. If $\mathbf{x}$ is in the domain $\mathbb{R}^n$, $T(\mathbf{x})$ denotes the image of $\mathbf{x}$. The set of images for all $\mathbf{x}$ in the domain is called the *range* of T.

To find the range of T, we want to know for what vectors $\mathbf{b}$ in $\mathbb{R}^m$ we can find a vector $\mathbf{x}$ in $\mathbb{R}^n$ such that $T(\mathbf{x}) = \mathbf{b}$. Since $T(\mathbf{x}) = A\mathbf{x}$, we want to know the set of all $\mathbf{b}$ in $\mathbb{R}^m$ for which we can find a vector $\mathbf{x}$ in $\mathbb{R}^n$ with $A\mathbf{x} = \mathbf{b}$. Thus, to determine the range of T, we need to find the set of all $\mathbf{b}$ in $\mathbb{R}^m$ for which the system of equations $A\mathbf{x} = \mathbf{b}$ has a solution (that is, we have Problem 1).

Recall also the following properties of functions, which we provide here again for convenience:

Definition 4.5.3 A function $T : \mathbb{R}^n \to \mathbb{R}^m$ is said to be *onto* if every $\mathbf{b}$ in $\mathbb{R}^m$ is the image of some $\mathbf{x}$ in the domain $\mathbb{R}^n$. That is, T is onto, if for every $\mathbf{b}$ in $\mathbb{R}^m$, there is an $\mathbf{x}$ in $\mathbb{R}^n$ such that

$$T(\mathbf{x}) = \mathbf{b}.$$

Since Theorem 4.5.1 tells us that we can always find a unique matrix A such that $T(\mathbf{x}) = A\mathbf{x}$, we see that a transformation T is onto if for every $\mathbf{b}$ in $\mathbb{R}^m$, there is a vector $\mathbf{x}$ in $\mathbb{R}^n$ such that

$$A\mathbf{x} = \mathbf{b}.$$

Thus, T is onto when the system of equations $A\mathbf{x} = \mathbf{b}$ has a solution for every $\mathbf{b}$ in $\mathbb{R}^m$ (that is, we have Problem 4).

Definition 4.5.4 A function $T : \mathbb{R}^n \to \mathbb{R}^m$ is said to be *one-to-one* if for any two vectors $\mathbf{x}_1$, $\mathbf{x}_2$ in the domain $\mathbb{R}^n$, $T(\mathbf{x}_1) = T(\mathbf{x}_2)$ implies $\mathbf{x}_1 = \mathbf{x}_2$.

Again, in view of Theorem 4.5.1, $T(\mathbf{x}_1) = T(\mathbf{x}_2)$ means that $A\mathbf{x}_1 = A\mathbf{x}_2$, so $A\mathbf{x}_1 - A\mathbf{x}_2 = A(\mathbf{x}_1 - \mathbf{x}_2) = \mathbf{0}$. To have this imply $\mathbf{x}_1 - \mathbf{x}_2 = \mathbf{0}$ is equivalent to $A\mathbf{x} = \mathbf{0}$ having only the solution $\mathbf{x} = 0$. Thus, T being one-to-one is equivalent to $A\mathbf{x} = \mathbf{0}$ having only the trivial solution $\mathbf{x} = \mathbf{0}$.

This shows that important properties of linear transformations can be studied by asking questions about the solution sets of the systems of linear equations $A\mathbf{x} = \mathbf{b}$ and $A\mathbf{x} = \mathbf{0}$ where A is the matrix representing the linear transformation. Since we have already answered those questions in Section 4.5.2, we just need to review that section to remind ourselves how to answer them.

We close this section with answering the question: So, where does the matrix A for a given linear transformation come from?

We show an example next, which will provide the main idea. The general case follows the same logic. The basic idea is that we want to see where the linear transformation maps the vectors from the standard basis.

■ **Example 4.24** Let T be a linear transformation from $\mathbb{R}^2$ to $\mathbb{R}^3$. Consider the images of the standard basis vectors $\mathbf{e}_1 = \begin{pmatrix} 1 \\ 0 \end{pmatrix}$ and $\mathbf{e}_2 = \begin{pmatrix} 0 \\ 1 \end{pmatrix}$ of $\mathbb{R}^2$. We will look at how T transforms $\mathbf{e}_1$ and $\mathbf{e}_2$. Let

$$T(\mathbf{e}_1) = T\begin{pmatrix} 1 \\ 0 \end{pmatrix} = \begin{pmatrix} a \\ b \\ c \end{pmatrix} \quad \text{and} \quad T(\mathbf{e}_2) = T\begin{pmatrix} 0 \\ 1 \end{pmatrix} \begin{pmatrix} d \\ e \\ f \end{pmatrix},$$

and A be the matrix with columns $T(\mathbf{e}_1)$ and $T(\mathbf{e}_2)$. That is,

$$A = \begin{pmatrix} a & d \\ b & e \\ c & f \end{pmatrix}.$$

Now, let's take a vector $\mathbf{x}$ in $\mathbb{R}^2$ and see how T transforms $\mathbf{x}$. If

$$\mathbf{x} = \begin{pmatrix} x_1 \\ x_2 \end{pmatrix} = x_1 \begin{pmatrix} 1 \\ 0 \end{pmatrix} + x_2 \begin{pmatrix} 0 \\ 1 \end{pmatrix},$$

then (using the linearity property of T from Definition 4.1.2)

$$T(\mathbf{x}) = T \begin{pmatrix} x_1 \\ x_2 \end{pmatrix} = x_1 T \begin{pmatrix} 1 \\ 0 \end{pmatrix} + x_2 T \begin{pmatrix} 0 \\ 1 \end{pmatrix} = x_1 \begin{pmatrix} a \\ b \\ c \end{pmatrix} + x_2 \begin{pmatrix} d \\ e \\ f \end{pmatrix}$$

$$= \begin{pmatrix} x_1 a + x_2 d \\ x_1 b + x_2 e \\ x_1 c + x_2 f \end{pmatrix}.$$

Now, let's find $A\mathbf{x}$:

$$A\mathbf{x} = A \begin{pmatrix} x_1 \\ x_2 \end{pmatrix} = \begin{pmatrix} a & d \\ b & e \\ c & f \end{pmatrix} \begin{pmatrix} x_1 \\ x_2 \end{pmatrix} = \begin{pmatrix} x_1 a + x_2 d \\ x_1 b + x_2 e \\ x_1 c + x_2 f \end{pmatrix}.$$

This shows that $T(\mathbf{x}) = A\mathbf{x}$. That is, *the linear transformation can be represented by multiplication of the matrix A, the columns of which are $T(\mathbf{e}_1)$ and $T(\mathbf{e}_2)$.* ∎

The proof of the general case can be done similarly. So, to construct the matrix A that represents a linear transformation $T : \mathbb{R}^m \to \mathbb{R}^n$, the columns of A will be the vectors $T(\mathbf{e}_1), T(\mathbf{e}_2), \ldots T(\mathbf{e}_m)$, where $\{\mathbf{e}_1, \mathbf{e}_2, \ldots, \mathbf{e}_m\}$ is the standard basis of $\mathbb{R}^m$.

Summary and What to Expect Next

The ideas and techniques we presented in this chapter form the foundation of linear algebra. If you were able to follow the examples and complete the exercises, you will likely transition with ease into your linear algebra class. In it, you will see rigorous proofs of the general facts that we chose to only illustrate with multiple examples here. You will gain a broader view of the subject and learn how to answer new questions. In many cases however, you will realize that those questions are equivalent to the fundamental problems we raised and answered in Section 4.5.2, just as we were able to answer, in Section 4.5.4, important questions about linear transformations. You will learn new matrix operations (e.g., finding the inverse of a matrix), new terms (e.g., rank, nullity, coordinate mapping), new mathematical objects and their properties (e.g., eigenvalues, eigenvectors, determinants, symmetric matrices), new techniques (e.g., diagonalization, singular value decomposition, principal component analysis), and more. All along though, you will keep coming back to the ideas and properties highlighted in this chapter. It may surprise you that when you study the vector space $\mathbb{R}^n$ you will be, in fact, learning every property of any other general vector space because it behaves mathematically just like $\mathbb{R}^n$. You will also learn that linear transformations between general vector spaces can be studied by considering the linear transformations from $\mathbb{R}^m$ to $\mathbb{R}^n$, which you saw in the previous section.

You have already learned a lot of linear algebra from these notes. It is our hope that you are now engaged and better prepared for what's to come next.

4.7 Suggested Videos and Further Reading

We highly recommend the excellent lessons on Linear Algebra from the 3Blue1Brown YouTube channel https://www.3blue1brown.com/topics/linear-algebra dedicated to "discovery and creativity in math." Grant Sanderson and his team have created an excellent collection of animated videos and written lessons that can be used to visualize many of the topics we have presented in this chapter, including vectors, span, matrix multiplication, column space, null space, rank, linear transformations, and general vector spaces. We recommend that you watch each video before reading the relevant topic in the chapter, then watch it again after you have completed the reading and worked through the examples and exercises in the section.

We also recommend the following textbooks for further reading. They are standard texts for undergraduate courses in linear algebra.

Bibliography

[1] Kirkwood, James R., and Bessie H. Kirkwood. *Linear Algebra.*. CRC Press, 2020.

[2] Lay, David C., Steven R. Lay, and Judi J. McDonald. *Linear Algebra and its applications*, 5th edition. Pearson, 2016.

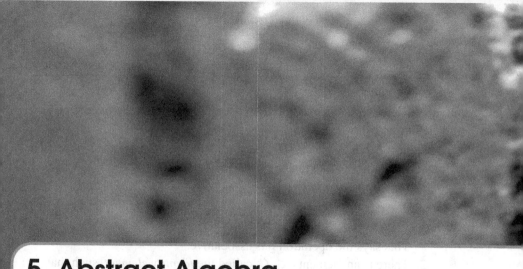

5. Abstract Algebra

The end of the 19th century and the beginning of the 20th century saw a shift in the methodology of mathematics. As examples from number theory, geometry, analysis, and algebraic equations came to be seen as special cases of general properties of certain abstract structures, formal definitions, such as those of groups, rings, and fields began to emerge. Questions about the structure and classification of these mathematical objects came to the forefront. Abstract algebra is the discipline that studies such general structures. The name was coined to distinguish it from elementary algebra, which uses variables to represent numbers in computation and reasoning.

In this chapter, we introduce the theory of algebraic groups. A group is a fundamental structure that is studied in most first courses in Abstract Algebra. The group structure can be found in many examples from mathematics, engineering, and science, and its versatility makes it an object of central interest.

Groups

We typically denote a group by G. A group has a set of elements and a binary operation defined on this set. The binary operation will most often be denoted by $\circ$, $*$, $+$, or juxtaposition. A binary operation describes how to combine two elements in the group to get another element in the group. Another way to say this is that a binary operation is a function $f : G \times G \to G$. If g_1 and g_2 are elements of G, and we denote the binary operation by $\circ$, then $g_1 \circ g_2$ is a

DOI: 10.1201/9781032623849-5

uniquely determined element of G. A group has additional structure given in the definition below.

Definition 5.1.1 A *group* $\langle G, \circ \rangle$ is a nonempty set of elements G with a binary operation $\circ$ such that

1. for every a, b, c in G,

$$a \circ (b \circ c) = (a \circ b) \circ c.$$

This says the operation $\circ$ is associative.

2. There is an element $e \in G$, called the *identity element*, such that

$$e \circ g = g \circ e = g$$

for every $g \in G$.

3. For each element $g \in G$, there is an element g^{-1}, called *the inverse* of g, such that

$$g \circ g^{-1} = g^{-1} \circ g = e.$$

Note that we must define the identity element before we define the inverse of an element.

We make the following observations:

1. It is not unusual for a student in their first abstract mathematics course to underestimate how important definitions are. A major factor in success in abstract mathematics is how precise your understanding of the definitions is.

2. If $+$ is the symbol for the binary operation, then it is customary to use $-g$ to represent the inverse of g and 0 to represent the identity element. We then have $g + 0 = 0 + g = g$, and $g + (-g) = (-g) + g = 0$.

3. In the statement $e + g = g + e$ we are not saying that $g_1 + g_2 = g_2 + g_1$ for every $g_1, g_2 \in G$. In the general case, the property is true only for the identity element. Also, in part 2 of the definition, the element e is independent of the element g, whereas in part 3 of the definition, the inverse of g depends on the choice of g.

We next prove some facts that are true for every group.

Theorem 5.1.1 The identity in a group is unique.

Proof. Recall, from Section 1.3.4, that to prove that something is unique, we typically assume there are two objects that satisfy the defining condition and show those two must be equal. Now, let $\langle G, \circ \rangle$ be a group. Suppose that e_1

and e_2 are two identities of $\langle G, \circ \rangle$. Since e_1 is an identity, then $e_1 \circ e_2 = e_2$ and since e_2 is an identity, then $e_1 \circ e_2 = e_1$. Thus, $e_1 = e_2$. ∎

Theorem 5.1.2 If $\langle G, \circ \rangle$ is a group, then the inverse of an element is unique.

Proof. Again, this can be proved by assuming that an element $g \in G$ has two inverses x and y, then showing that $x = y$. So, let x and y be two such inverses for g. This means that $x \circ g = e$ and $g \circ y = e$. Now, using the properties from Definition 5.1.1, we have

$$y = y \circ e = e \circ y = (x \circ g) \circ y = x \circ (g \circ y) = x \circ e = x.$$

This shows that $x = y$, and the inverse of any element of the group is unique.
∎

Definition 5.1.2 If the binary operation on a group is commutative, that is if $g_1 \circ g_2 = g_2 \circ g_1$ for all $g_1, g_2 \in G$, then the group is said to be *abelian*.

When a group G is abelian, it is common to refer to the group operation of G as $+$. So, from now on, we will use additive notation for abelian groups.

Definition 5.1.3 If a group G has a finite number of elements, G is called a *finite group*. We use $|G|$ to denote the number of elements in G.

Exercise 5.1 Consider the set of integers $\mathbb{Z}$. Is the binary operation on $\mathbb{Z}$, defined by $a \circ b = a^2 + b^2$ commutative? Is it associative? ∎

Exercise 5.2 Consider the set of integers $\mathbb{Z}$. Is the binary operation on $\mathbb{Z}$, defined by $a * b = a^2 + b^2 - ab$ commutative? Is it associative? ∎

Exercise 5.3 If $\langle G, \circ \rangle$ is a group and $a, b, c \in G$ with $(a \circ b) = (a \circ c)$, show that $b = c$. ∎

Exercise 5.4 If $\langle G, \circ \rangle$ is a group and $a, b, c \in G$ with $(c \circ a) = (c \circ b)$, show that $a = b$. ∎

Exercise 5.5 Show that $\langle G, \circ \rangle$ is abelian if and only if $(a \circ b)^2 = (a^2 \circ b^2)$ for all $a, b \in G$. ∎

Exercise 5.6 If $\langle G, \circ \rangle$ is a group and $a, b \in G$, show that
1. $(a^{-1})^{-1} = a$,
2. $(a \circ b)^{-1} = b^{-1} \circ a^{-1}$.

5.2 Examples of Groups

You are already familiar with two sets · each with a binary operation that makes them groups – the set of real numbers $\mathbb{R}$ and the set of all integers $\mathbb{Z}$, which we present in the first two examples below. In fact, the definition of a group has emerged as their generalization.

■ **Example 5.1** $\langle \mathbb{R} - \{0\}, \cdot \rangle$: The set of nonzero real numbers $\mathbb{R} - \{0\}$, with the operation given by the usual multiplication of numbers. The conditions in Definition 5.1.1 are satisfied, as we have:
1. For any $a, b, c \in \mathbb{R}$, $(ab)c = a(bc)$, as we know that multiplication of real numbers is associative;
2. The identity element is the number 1, as we know $1 \cdot a = a \cdot 1 = a$;
3. The inverse of each element $a \in \mathbb{R} - \{0\}$ is its reciprocal $\frac{1}{a}$, because $a \neq 0$ and $a \cdot \frac{1}{a} = 1$.
■

■ **Example 5.2** $\langle \mathbb{Z}, + \rangle$: The set of integers $\mathbb{Z}$ with the addition operation $+$. Checking the conditions from the definition of a group, we use the known properties of addition:
1. It is associative, because for any $a, b, c \in \mathbb{Z}$, $(a+b)+c = a+(b+c) = a+b+c$;
2. The number 0 is the identity element, since for every integer a, $a+0 = 0+a = a$.
3. The inverse of $a \in Z$, is $-a \in \mathbb{Z}$, because $a+(-a) = (-a)+a = 0$.
■

■ **Example 5.3** $\langle \mathbb{Z}_n, + \rangle$: The set of integers modulo n with $(\bmod\ n)$ addition (see Section 6.7 of the Appendix on modular arithmetic). The set of elements is $\mathbb{Z}_n = \{0, 1, 2, 3, \ldots, n-1\}$ and the binary operation $(\bmod\ n)$ addition (which we denote here by $\oplus$) is defined by

$$g_1 \oplus g_2 = g_1 + g_2 \ (\bmod\ n), \quad \text{for } g_1, g_2 \in \mathbb{Z}.$$

The group $\mathbb{Z}_n$ is a finite group with n elements, and we write $|\mathbb{Z}_n| = n$.
1. The associativity property of $\oplus$ follows from the associativity of adding integers and from Theorem 6.7.2 in the Appendix.

2. The identity element is 0, as we have, for any $a \in \mathbb{Z}_n$,

$$a \oplus 0 = (a+0) \ (\text{mod} \ n) = a,$$

and, similarly, $0 \oplus a = a$.

3. If $a \in \mathbb{Z}_n$, its inverse is $n - a$, since

$$a \oplus (n-a) = a + (n-a) \ (\text{mod} \ n) = n \ (\text{mod} \ n) = 0.$$

Similarly, $(n-a) \oplus a = 0$.

As an example, for $n = 6$, $|\mathbb{Z}_6| = 6$, the identity in $\mathbb{Z}_6$ is 0, and the inverse of 4 is 2.   ∎

∎ **Example 5.4** $\langle \mathbb{Q}_+, \cdot \rangle$: The set of positive rational numbers (denoted $\mathbb{Q}_+$) with the regular multiplication operation.

This is very similar to Example 5.1, but we only consider the set of positive fractions. The identity element is the number 1, and the inverse of a fraction $\frac{p}{q} \in \mathbb{Q}_+, q \neq 0$, is its reciprocal $\frac{q}{p}$.   ∎

Once we understand how the operation on a group G is defined, it may become cumbersome to use the $\circ$ notation in expressions. It is very common to omit the symbol and write ab in place of $a \circ b$, much as we do with regular multiplication. So, from now on, we will simply keep in mind that G is a group with a group operation, for which we omit the symbol. We will write $(ab)c = a(bc)$ for the associativity property, $aa^{-1} = e$, $ea = a$, and so on.

> **Exercise 5.7** Prove that the set of 2×2 matrices with nonzero determinants is a group under matrix multiplication.
>
> 1. What is the identity of the group?
> 2. What is the inverse of $\begin{pmatrix} a & b \\ c & d \end{pmatrix}$?
>
>   ∎

> **Exercise 5.8** Prove that the set of 2×2 matrices with determinants equal to 1 is a group under matrix multiplication.   ∎

> **Exercise 5.9** Prove that the set of 2×2 diagonal matrices with nonzero diagonal entries forms a group under matrix multiplication.   ∎

Exercise 5.10 Let G be a group. Prove that if two elements $a, b \in G$ commute (that is, $ab = ba$), then their inverses a^{-1} and b^{-1} also commute.

■

Exercise 5.11 If G is a finite group, prove that there exists a positive integer N, such that $a^N = e$ for all $a \in G$.   ■

Exercise 5.12 Let G be a group, and let $x, y, a \in G$. Prove that if $xay = a^{-1}$, then $yax = a^{-1}$.   ■

Exercise 5.13 Show that if every element in a group is its own inverse, the group must be abelian. *Hint.* If each element of a group G is its own inverse, we have $a^{-1}a = aa = e$ for all $a \in G$. Because for any $a, b \in G$, we know that $ab \in G$, we have $abab = e$. Now multiply this from the left and from the right by an appropriate inverse to obtain $ab = ba$.   ■

5.3 The Symmetric Group S_n

In this section, we will focus more closely of one group that is of special interest in abstract algebra and in various branches of mathematics – the symmetric group S_n of n elements. The group S_n is also of interest in physics, engineering, and chemistry.

For a positive integer n, we consider the group of permutations on $\{1, 2, \ldots, n\}$; that is, the set of all one-to-one and onto functions (bijections) from the set $G = \{1, 2, \ldots, n\}$ to itself. As our examples below show, it is convenient to express the bijections f as sequences.

■ **Example 5.5** Consider for example S_4 with the function f on $\{1, 2, 3, 4\}$ defined by $f(1) = 3$, $f(2) = 1$, $f(3) = 4$, and $f(4) = 2$. This function, thus, takes $1 \rightarrow 3$, $3 \rightarrow 4$, $4 \rightarrow 2$, and $2 \rightarrow 1$. We can also express this as

$$f = \begin{pmatrix} 1 & 2 & 3 & 4 \\ 3 & 1 & 4 & 2 \end{pmatrix},$$

or, equivalently, as $(1\ 3\ 4\ 2)$, with the understanding that each number maps to the number on its right and the last number in the sequence maps to the first.

■

■ **Example 5.6** Consider now S_5 and the permutation f in S_5 defined as: $f(1) = 5$, $f(2) = 3$, $f(3) = 4$, $f(4) = 1$, and $f(5) = 2$. We can also represent this as $1 \to 5, 5 \to 2, 2 \to 3, 3 \to 4, 4 \to 1$, or we can write

$$f = \begin{pmatrix} 1 & 2 & 3 & 4 & 5 \\ 5 & 3 & 4 & 1 & 2 \end{pmatrix}.$$

In this case, the sequence that represents f is $(1\ 5\ 2\ 3\ 4)$.

■

We turn the set of permutations into a group using compositions of functions. Recall that if we look at two functions $f(x)$ and $g(x)$ defined on the real line their composition $f \circ g(x)$ is defined as $f \circ g(x) = f(g(x))$. For example, if $f(x) = x^2 + 1$ and $g(x) = 3x$, then $f \circ g(x) = f(g(x)) = f(3x) = (3x)^2 + 1$. That is, we apply the function on the right first. Because of that, $f \circ g(x)$ is often read "f after g" (see Also Section 6.10 of the Appendix).

■ **Example 5.7** On S_3, let

$$f = \begin{pmatrix} 1 & 2 & 3 \\ 3 & 2 & 1 \end{pmatrix}, \text{ and } g = \begin{pmatrix} 1 & 2 & 3 \\ 2 & 3 & 1 \end{pmatrix},$$

so $f(1) = 3, f(2) = 2, f(3) = 1$, and $g(1) = 2, g(2) = 3, g(3) = 1$. Now

$$\begin{array}{ccccc} f \circ g(1) & = & f(g(1)) & = & f(2) = 2 \\ f \circ g(2) & = & f(g(2)) & = & f(3) = 1 \\ f \circ g(3) & = & f(g(3)) & = & f(1) = 3 \end{array}.$$

So, we can write

$$f \circ g = \begin{pmatrix} 1 & 2 & 3 \\ 2 & 1 & 3 \end{pmatrix}.$$

■

We are now ready to give the following definition.

Definition 5.3.1 The *symmetric group* S_n defined over any set G of n objects, is the set of all one-to-one and onto functions (bijections) $f : G \to G$ with the operation composition of functions.

In our last example, the function $f \circ g(x)$ mapped 3 to itself (and the function $f(x)$ mapped 2 to itself). In this case, the cycle notation we used in Example 5.6 is not possible. Instead, we can write

$$f \circ g(x) = (1\ 2)(3)$$

to indicate the mapping given by the cycle $(1\ 2)$ and that 3 is mapped to itself. This is a very useful observation, as it allows us to represent any permutation into a product of disjoint cycles, as we will see next.

■ **Example 5.8** Consider the permutation

$$f = \begin{pmatrix} 1 & 2 & 3 & 4 & 5 & 6 & 7 & 8 & 9 \\ 9 & 1 & 5 & 3 & 4 & 7 & 6 & 8 & 2 \end{pmatrix}.$$

To represent a permutation as a product of cycles, one chooses a number (most often 1) and traces the path that the permutation defines until you return to the chosen number (without repeating the first number). So our first cycle for this permutation could be $(1\ 9\ 2)$.

To get the subsequent cycle, choose a number that does not appear in the first cycle. It is often handy to cross out the numbers that have been used and start the second cycle with the smallest number that has not previously been used. In this case that is the number 3. So our next cycle would be $(3\ 5\ 4)$. Continuing the process yields $(6\ 7)$ and (8). The representation of this permutation into cycles is $f = (1\ 9\ 2)(3\ 5\ 4)(6\ 7)(8)$.   ■

Example 5.8 illustrates a general method of decomposing a permutation into disjoint cycles. In this representation each number appears in exactly one cycle. Note that the order in which we list the cycles is not important. We can express the factorization in different orders.

Cycles can be further "broken down" into transpositions. A *transposition* is a cycle of two elements. We will first illustrate this with an example.

■ **Example 5.9** We will show that $(2\ 7\ 4\ 5)$ can be written as the following product of transpositions:

$$(2\ 7\ 4\ 5) = (2\ 5)(2\ 4)(2\ 7).$$

We will trace the path of each element of the cycle through the product of transpositions $(2\ 5)(2\ 4)(2\ 7)$ to verify it generates $(2\ 7\ 4\ 5)$. Remember we start on the right when we do function composition, so, for $(2\ 5)(2\ 4)(2\ 7)$, we do $(2\ 7)$ first and then proceed right to left.

Starting with 2, we have the transposition $(2\ 7)$ that tells us $2 \to 7$. Moving to the left, the transpositions $(2\ 4)$ and $(2\ 5)$ do nothing to 7, so the composition of transpositions gives us $2 \to 7$.

Next, we take 7. The transposition $(2\ 7)$ moves $7 \to 2$, and $(2\ 4)$, tells us that $2 \to 4$. The transposition $(2\ 5)$ does nothing to 4 and, as a result, we have that $7 \to 4$.

Continuing with 4, the transposition $(2\ 7)$ does not involve 4, but $(2\ 4)$ moves $4 \to 2$, and $(2\ 5)$ moves $2 \to 5$. The result then is that $4 \to 5$.

Finally for 5, $(2\ 7)$ does not involve 5 and neither does $(2\ 4)$. Applying $(2\ 5)$ gives $5 \to 2$.

This proves that $(2\ 7\ 4\ 5) = (2\ 5)(2\ 4)(2\ 7)$.

∎

We need to stress that when we compose transpositions, the order in which they act *does* matter. Check, for example, that $(2\ 7\ 4\ 5) \neq (2\ 4)(2\ 5)(2\ 7)$. This is in contrast with our earlier observation that when a permutation is expressed as a product of disjoint cycles, the order in which we write the cycles doesn't matter.

∎ **Example 5.10** Show that $(1\ 2\ 3\ 4\ 5) = (1\ 5)(1\ 4)(1\ 3)(1\ 2)$.

The rightmost transposition moves $1 \to 2$, and the rest of the transpositions do not move 2. For 2, $(1\ 2)$ moves $2 \to 1$, then $(1\ 3)$ involves 1 (it moves 1 to 3), so we have that $2 \to 3$. Thus, $1 \to 3$. None of the transpositions to the left of $(1\ 3)$ involves 3, so we have $2 \to 3$. Similar arguments show that $3 \to 4$ and that $4 \to 5$. Finally, 5 is moved only by the leftmost transposition, which gives $5 \to 1$.

∎

Notice the pattern in these two examples. This is not a coincidence, and we now give the general rule.

Theorem 5.3.1 Any cycle $(a_1\ a_2\ \ldots\ a_n)$ can be written as a product of transpositions in the following way:

$$(a_1\ a_2\ \ldots\ a_n) = (a_1\ a_n)(a_1\ a_{n-1})\ldots(a_1\ a_2). \tag{5.1}$$

Proof. Note that in the composition $(a_1\ a_2\ \ldots\ a_n) = (a_1\ a_n)(a_1\ a_{n-1}) \ldots(a_1\ a_{k+1})(a_1\ a_k)\ldots(a_1\ a_2)$, the first and the last element of the cycle are moved by a single transposition – $(a_1\ a_2)$ and $(a_1\ a_n)$, respectively. Every other element a_k, $1 < k < n$ is moved by exactly two transpositions – the transposition $(a_1\ a_k)$ and the preceding transposition $(a_1\ a_{k+1})$. The product of these transpositions means $a_k \to a_1$ and $a_1 \to a_{k+1}$, ensuring that $a_k \to a_{k+1}$ for each k, $1 < k < n$. Thus $(a_1\ a_n)(a_1\ a_{n-1})\ldots(a_1\ a_2) = (a_1\ a_2\ \ldots\ a_n)$. ∎

Theorem 5.3.1 gives a convenient rule for decomposing a cycle into a product of transpositions. However, this decomposition is not unique. There are many other ways to write the same permutation. To illustrate this, note that the permutation from Example 5.10 can be written as

$$(1\ 2\ 3\ 4\ 5) = (1\ 5)(1\ 4)(1\ 3)(1\ 2),$$
$$(1\ 2\ 3\ 4\ 5) = (5\ 4)(5\ 3)(5\ 2)(5\ 1),$$
$$(1\ 2\ 3\ 4\ 5) = (5\ 4)(5\ 2)(5\ 1)(1\ 4)(3\ 2)(4\ 1),$$

and in many other ways (check!).

These examples show that not only the decomposition into transpositions in not unique, but also that the various decompositions don't even have to have the same number of transpositions. Interestingly, one can show that even though the decomposition is not unique, the number of transpositions involved in the various decompositions is *always even* or *always odd* for any given cycle. This further shows that every permutation in S_n can be decomposed either into an even number of transpositions (and such permutations are called *even*) or an odd number of transpositions (and such permutations are called *odd*). This separation of the elements of S_n into even and odd plays an important role in group theory and in abstract algebra, in general.

Exercise 5.14 Write the following permutations as a product of disjoint cycles.

1. $\begin{pmatrix} 1 & 2 & 3 & 4 & 5 \\ 2 & 1 & 3 & 5 & 4 \end{pmatrix}$;

2. $\begin{pmatrix} 1 & 2 & 3 & 4 & 5 \\ 5 & 4 & 1 & 3 & 2 \end{pmatrix}$;

3. $\begin{pmatrix} 1 & 2 & 3 & 4 & 5 \\ 1 & 3 & 2 & 5 & 4 \end{pmatrix}$.

Exercise 5.15 Decompose the following cycles into transpositions.

1. $(5\ 8\ 2\ 4)$;
2. $(8\ 3\ 6\ 1\ 2)$;
3. $(2\ 7\ 8\ 5\ 4\ 3)$.

Exercise 5.16 Express the following composition of transpositions as a single cycle.

1. $(1\ 6)(1\ 5)(1\ 4)(1\ 3)(1\ 2)$;
2. $(3\ 1)(4\ 3)(6\ 3)(5\ 1)(2\ 3)$;
3. $(5\ 2)(6\ 2)(9\ 2)(3\ 2)$.

Expressing a Binary Operation as a Table

We often specify the binary operation for groups with only few elements by a table, called a *Cayley table*. The idea is the same as for the multiplication table of integers that you learn in grade school. Our example below is the multiplication table for the symmetric group on three symbols S_3. Remember that the elements of S_3 are permutations and the multiplication operation is composition of functions. The group S_3 is an important group for a number of reasons. One reason is that it is the smallest nonabelian group. Another is that it models the rigid motions of an equilateral triangle. We will come back to this later.

The group S_3 has six elements (the number of permutations of three elements). Let

$$\rho_0 = \begin{pmatrix} 1 & 2 & 3 \\ 1 & 2 & 3 \end{pmatrix}, \quad \rho_1 = \begin{pmatrix} 1 & 2 & 3 \\ 2 & 3 & 1 \end{pmatrix}, \quad \rho_2 = \begin{pmatrix} 1 & 2 & 3 \\ 3 & 1 & 2 \end{pmatrix}.$$

and

$$\mu_1 = \begin{pmatrix} 1 & 2 & 3 \\ 1 & 3 & 2 \end{pmatrix}, \quad \mu_2 = \begin{pmatrix} 1 & 2 & 3 \\ 3 & 2 & 1 \end{pmatrix}, \quad \mu_3 = \begin{pmatrix} 1 & 2 & 3 \\ 2 & 1 & 3 \end{pmatrix},$$

To fill in the Cayley table, we need to compute all pairwise products. We give two as examples, the rest are left as an exercise.

■ **Example 5.11** We will show that $\rho_2 \circ \mu_3 = \mu_1$ and that $\rho_2 \circ \mu_1 = \mu_2$. All other products can be computed similarly.

Fo the first product, we have

$$\begin{array}{rclcl} \rho_2 \circ \mu_3(1) & = & \rho_2(\mu_3(1)) & = & \rho_2(2) = 1 \\ \rho_2 \circ \mu_3(2) & = & \rho_2(\mu_3(2)) & = & \rho_2(1) = 3 \\ \rho_2 \circ \mu_3(3) & = & \rho_2(\mu_3(3)) & = & \rho_2(3) = 2 \end{array}$$

This shows that

$$\rho_2 \circ \mu_3 = \begin{pmatrix} 1 & 2 & 3 \\ 1 & 3 & 2 \end{pmatrix} = \mu_1.$$

For the second product, we compute

$$
\begin{aligned}
\rho_2 \circ \mu_1(1) &= \rho_2(\mu_1(1)) &= \rho_2(1) = 3 \\
\rho_2 \circ \mu_1(2) &= \rho_2(\mu_1(2)) &= \rho_2(3) = 2 \\
\rho_2 \circ \mu_1(3) &= \rho_2(\mu_1(3)) &= \rho_2(2) = 1
\end{aligned}
$$

which shows that

$$
\rho_2 \circ \mu_1 = \begin{pmatrix} 1 & 2 & 3 \\ 3 & 2 & 1 \end{pmatrix} = \mu_2.
$$

∎

The complete multiplication table (Cayley table) is given in Table 5.1. To compute the product $a \circ b$ for any two elements a and b of S_3, take intersection of the row having b at the left and the column having a at the top. So, $\mu_1 \circ \mu_2 = \rho_1$, $\mu_2 \circ \rho_1 = \mu_3$, and so on.

$\circ$	ρ_0	ρ_1	ρ_2	μ_1	μ_2	μ_3
ρ_0	ρ_0	ρ_1	ρ_2	μ_1	μ_2	μ_3
ρ_1	ρ_1	ρ_2	ρ_0	μ_2	μ_3	μ_1
ρ_2	ρ_2	ρ_0	ρ_1	μ_3	μ_1	μ_2
μ_1	μ_1	μ_3	μ_2	ρ_0	ρ_2	ρ_1
μ_2	μ_2	μ_1	μ_3	ρ_1	ρ_0	ρ_2
μ_3	μ_3	μ_2	μ_1	ρ_2	ρ_1	ρ_0

Table 5.1: The multiplication table (Cayley table) for the symmetric group S_3.

This table demonstrates some properties that all such tables must have including that each row contains every element of the group as does each column.

Note that the table for S_3 shows that the group is not abelian – for instance $\mu_2 \circ \rho_1 = \mu_3$ and $\rho_1 \circ \mu_2 = \mu_1$, and we know $\mu_1 \neq \mu_3$. Thus S_3 is not abelian.

5.3.2 Rotations and Reflections of Polyhedra

Another convenient way to think of the elements of S_n is as rotations and reflections of regular polyhedra. We expand on this explanation for S_3.

∎ **Example 5.12** The symmetric group S_3 gives all possible ways in which an equilateral triangle can be repositioned to lie on itself.

Consider an equilateral triangle with vertices numbered 1, 2, and 3 (Figure 5.1).

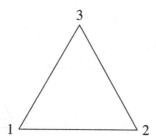

Figure 5.1: How many different ways are there to reposition an equilateral triangle to lie on itself? This is the question from which S_3 originated.

We consider all possible counterclockwise rotations and all reflections that reposition the vertices in a way that maps the triangle onto itself. We refer to those mappings as *symmetries of the triangle*. There are six possible mappings of the triangle onto itself. Figures 5.2 and 5.3 visualize the symmetries μ_3 and ρ_1.

$$\rho_0 = \begin{pmatrix} 1 & 2 & 3 \\ 1 & 2 & 3 \end{pmatrix} - \text{the identity;}$$

$$\rho_1 = \begin{pmatrix} 1 & 2 & 3 \\ 2 & 3 & 1 \end{pmatrix} - \text{rotate counterclockwise } 120°;$$

$$\rho_2 = \begin{pmatrix} 1 & 2 & 3 \\ 3 & 1 & 2 \end{pmatrix} - \text{rotate counterclockwise } 240°;$$

$$\mu_1 = \begin{pmatrix} 1 & 2 & 3 \\ 1 & 3 & 2 \end{pmatrix} - \text{reflect over vertex 1;}$$

$$\mu_2 = \begin{pmatrix} 1 & 2 & 3 \\ 3 & 2 & 1 \end{pmatrix} - \text{reflect over vertex 2;}$$

$$\mu_3 = \begin{pmatrix} 1 & 2 & 3 \\ 2 & 1 & 3 \end{pmatrix} - \text{reflect over vertex 3.}$$

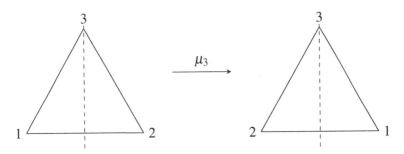

Figure 5.2: The symmetry μ_3 is a reflection over vertex 3.

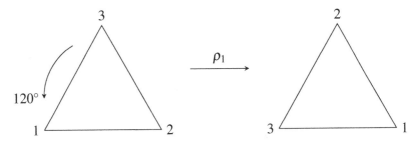

Figure 5.3: The symmetry ρ_1 is a rotation by $120°$ counterclockwise.

The operation $\circ$ in S_3 can be interpreted as performing rigid rotations/reflections of an equilateral triangle in a designated order. For example, $\mu_1 \circ \rho_2$ is a counterclockwise $240°$ rotation first, followed by a reflection over vertex 1. This accomplishes the same as reflecting the rectangle over vertex 3, so $\mu_1 \circ \rho_2 = \mu_3$. With $\circ$ denoting the composition of any two rotation/reflection operations, we leave it to you as an exercise to show that Table 5.1 gives us the results from all such compositions. Once again, we stress that the order of operations matters: rotating counterclockwise $240°$ and then reflecting over vertex 1 ($\mu_1 \circ \rho_2$) does not produce the same result as reflecting over vertex 1 first, followed by $240°$ rotation ($\rho_2 \circ \mu_1$). Said another way, S_3 is not abelian.   ■

Exercise 5.17 Consider the symmetric group S_4. Just as S_3 can be thought of as the symmetry group of an equilateral triangle, S_4 can be interpreted as the symmetry group of a square with vertices numbered 1, 2, 3, and 4.
   1. List all symmetries of the square. For example, $f = (1\ 2\ 3\ 4)$ is a $90°$ counterclockwise rotation, while $g = (1\ 2)(3\ 4)$ is a reflection about the vertical axis of symmetry (see Figures 5.4 and 5.5).

> 2. Give the Cayley table for S_4.
> 3. Is S_4 abelian?   ∎

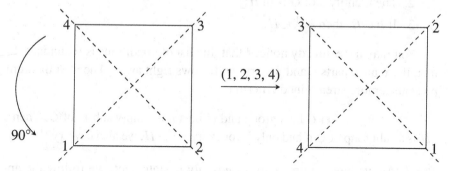

Figure 5.4: The symmetry $(1,2,3,4)$ is a 90° counterclockwise rotation.

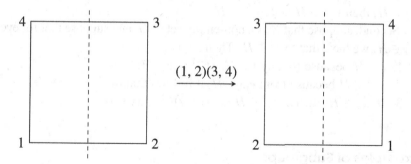

Figure 5.5: The symmetry $(1,2)(3,4)$ is a reflection about the vertical axis of symmetry.

Subgroups

> **Definition 5.4.1 — Subgroup.** If $\langle G, \circ \rangle$ is a group and H is a subset of G, we say that H is a *subgroup* of G if $\langle H, \circ \rangle$ is itself a group.

Notice that the binary operation in the subgroup is the same as in the group G.

Not every subset of a group is a group. For example, $H = \{\pm 1, \pm 3, \pm 5, \pm 7, \ldots\}$ is not a subgroup of $\mathbb{Z}, +\rangle$ because it doesn't contain the identity 0, nor is it closed under addition: e.g., $1 + 3 = 4 \notin H$.

Our first task is to find ways to determine if a subset $H \subseteq G$ is a subgroup. Theorems 5.4.1 and 5.4.2 provide some criteria.

> **Theorem 5.4.1** A subset $H \subseteq G$ is a subgroup of G if and only if all of the following conditions are satisfied:
>   1. The binary operation on G is closed on H;
>   2. The identity e of G is in H;
>   3. If $h \in H$, then $h^{-1} \in H$.

You may have already noticed that this list of conditions is redundant. In fact, if we have parts 1 and 3, part two follows right away. The next theorem presents a more streamlined criterion.

> **Theorem 5.4.2** Let G be a group and H be a non-empty subset of G. Then H is a subgroup of G if and only if for every $x, y \in H$, we also have $xy^{-1} \in H$.

Proof. As we have to prove an *if-and-only-if* statement, we follow the approach from Section 1.3.6, to present the two relevant *if-then* proofs.

First, let H be a subgroup of G. Then, by definition, H is a group. So, if $x, y \in H$, then $y^{-1} \in H$ and $xy^{-1} \in H$.

Second, suppose that H is a non-empty set of G, and suppose that for every $x, y \in H$, we have that $xy^{-1} \in H$. Then,
   1. $e \in H$ because for any $x \in H$, $xx^{-1} = e \in H$.
   2. $x^{-1} \in H$ because from $e, x \in H$, it follows that $ex^{-1} = x^{-1} \in H$.
   3. If $x, y \in H$, then $x, y^{-1} \in H$, so $x(y^{-1})^{-1} = xy \in H$.

∎

5.4.1   Examples of Subgroups

For several of the examples here and those later in the chapter, you may need to review some background material on integer division and modular arithmetic presented in Section 6.7 of the Appendix. There, you will find the relevant definitions and theorems, an introduction to the notation $m\mathbb{Z} + r$ $(m, r \in \mathbb{Z})$, and numerous examples.

For the examples that follow, we will use Theorem 5.4.2 to verify they are subgroups.

■ **Example 5.13** Consider $2\mathbb{Z} = \langle 2\mathbb{Z}, + \rangle$, the set of even integers with the operation +. Then $\langle 2\mathbb{Z}, + \rangle$ is a subgroup of $\langle \mathbb{Z}, + \rangle$.

If $a, b \in 2\mathbb{Z}$, then $a - b$ is a difference of two even numbers, which is even (See Exercise 1.18). Thus, this is a subgroup.   ■

■ **Example 5.14** Consider again the group $\langle \mathbb{Z}, + \rangle$.

- Let $3\mathbb{Z} = \{\dots, -6, -3, 0, 3, 6, 9, \dots\}$, that is, $3\mathbb{Z}$ is the set of all integers divisible by 3. Then $\langle 3\mathbb{Z}, + \rangle$ is a subgroup of $\langle \mathbb{Z}, + \rangle$.
- In general, if $n \geq 2$ is an integer, then the subset $n\mathbb{Z}$ of all integers that are divisible by n is a subgroup of $\langle \mathbb{Z}, + \rangle$.

To see this, let $a, b \in \mathbb{Z}$. This means $a = nk$ for some $k \in \mathbb{Z}$ and $b = nm$ for some $m \in \mathbb{Z}$. Then $a - b = nk - nm = n(k - m)$. Since $k - m \in \mathbb{Z}$, this proves that $n\mathbb{Z}$ is a subgroup of $\langle \mathbb{Z}, + \rangle$.   ■

■ **Example 5.15** Consider the group G of all 2×2 matrices and nonzero determinants from Exercise 5.7. The operation is matrix multiplication. With this, the set

$$H = \left\{ \begin{pmatrix} 1 & n \\ 0 & 1 \end{pmatrix} \,\middle|\, n \in \mathbb{Z} \right\}$$

is a subgroup of the group G.

To verify that H is a subgroup of G, let $A, B \in H$. This means that

$$A = \begin{pmatrix} 1 & k \\ 0 & 1 \end{pmatrix}, \quad B = \begin{pmatrix} 1 & m \\ 0 & 1 \end{pmatrix} \quad \text{for some } k, m \in \mathbb{Z}.$$

Because (see Theorem 4.2.2)

$$B^{-1} = \begin{pmatrix} 1 & -m \\ 0 & 1 \end{pmatrix},$$

we have

$$AB^{-1} = \begin{pmatrix} 1 & k \\ 0 & 1 \end{pmatrix} \begin{pmatrix} 1 & -m \\ 0 & 1 \end{pmatrix} = \begin{pmatrix} 1 & k-m \\ 0 & 1 \end{pmatrix} \in H, \text{ since } k - m \in \mathbb{Z}.$$

Thus, H is a subgroup.

                                                                                    ■

■ **Example 5.16** Let G be a group and $g \in G$. Consider the set H of all integer powers of g: $H = \{g^n \mid n \in \mathbb{Z}\}$. Then H is a subgroup of G.

To see this, let $a, b \in G$. Then $a = g^k$ and $b = g^m$, for some $k, m \in \mathbb{Z}$. Now,

$$ab^{-1} = (g^k)(g^m)^{-1} = (g^k)(g^{-m}) = g^{k-m} \in H, \text{ since } k - m \in \mathbb{Z}.$$

Thus, H is a subgroup of G. In fact, H is the smallest subgroup of G that contains g.

                                                                                    ■

■ **Example 5.17** Consider the symmetric group S_n, and recall that each permutation in S_n can be written as a product of transpositions. Although this representation is not unique, the number of transpositions in the product is either always even or always odd. This leads to separating the elements of S_n into even permutations and odd permutations. The subset $A_n \subset S_n$ of even permutations is a subgroup of S_n called the *alternating group* on the set $\{1, 2, \ldots, n\}$. The key idea here, even though it is somewhat technical to prove, is that the identity permutation is even (see Exercise 5.21).   ■

Exercise 5.18 Prove that if H_1 and H_2 are subgroups of G, then $H_1 \cap H_2$ is a subgroup. Is this true for $H_1 \cup H_2$?   ▪

Exercise 5.19 Let a be a fixed element in a group G. Prove that the set $H = \{x \in G \mid ax = xa\}$ is a subgroup of G.   ▪

Exercise 5.20 Let G be a group and $g \in G$. Prove that $H = \{g^n \mid n \in \mathbb{Z}\}$. Prove that H is the smallest subgroup of H that contains g (see Example 5.16, where we proved that H is a subgroup of G).   ▪

Exercise 5.21 Consider the alternating subgroup group A_n of the symmetric group S_n (see Example 5.17). Assume you have already proved that the identity permutation is even. Show that A_n is a subgroup of S_n.   ▪

Exercise 5.22 Consider the group G of all $n \times n$ matrices with nonzero determinant with the operation matrix multiplication. Let $H \subset G$ be the subset of all diagonal matrices. Show that H is a subgroup of G.   ▪

Exercise 5.23 Let G be a group and n be a fixed integer. Consider the set

$$H = \{x \in G \mid x^n = e\}.$$

Show that H is a subgroup.   ▪

5.5 Equivalence Relations

There are some special subgroups that are closely related to relations defined on sets. We explore this connection next.

Let A be a non-empty set. For what we present in this section, A does not have to have any structure, it can be any set.

Definition 5.5.1 For a set A, a *relation* R on A is any subset R of $A \times A$. For any $(a, b) \in A \times A$, either (a, b) is in the subset R or not. If $(a, b) \in R$, we say that a is *R-related* to b and we will write this as aRb.

Our focus will be on a special class of relations, called equivalence relations.

Definition 5.5.2 We say that R is an *equivalence relation* on a set A if the following conditions are satisfied for all $a, b, c \in A$:

1. aRa (R is reflexive);
2. If aRb, then bRa (R is symmetric);
3. If aRb and bRc, then aRc (R is transitive).

We give a few examples here and more examples in the exercises.

Examples of Relations

■ **Example 5.18** Let A be the set of people in a certain town and R is "in the same family" relation. That is, for any two people $a, b \in A$, aRb means that a is in the same family as b. Then R is an equivalence relation on A because all three conditions from Definition 5.5.2 are satisfied: (1) You are in the same family as yourself; (2) If two people a and b are in the same family, then surely b and a are in the same family, and (3) If person a is in the same family as person b, and person b is in the same family as person c, then a and c are in that same family.   ■

■ **Example 5.19** Let A be the set of people in a certain town and R is the "is older than" relation. That is, aRb means that person a is older than person b. This is not an equivalence relation because, e.g., the symmetric condition fails.   ■

■ **Example 5.20** Consider the set of integers $\mathbb{Z}$, a positive nonzero integer n and the "congruence (mod n)" relation R. Recall (Section 6.7 of the Appendix) that a is congruent to b (mod n) (that is aRb) if $a - b = np$ for some $p \in \mathbb{Z}$. We show this is an equivalent relation. Let $a, b, c \in \mathbb{Z}$.

1. We have $a - a = 0$, for any $a \in \mathbb{Z}$, so the relation is reflexive;
2. If $a, b \in \mathbb{Z}$ and $a - b = pn$ for some integer p, then $b - a = -np = n(-p)$, so the relation is reflexive;
3. If $a, b, c \in \mathbb{Z}$ satisfy $a - b = np$ for some integer p and $b - c = nq$ for some integer q, then $a - c = (a - b) + (b - c) = np + nq = n(p + q)$, so R is transitive.   ■

As our next example shows, equivalence relations arise naturally in the context of set partitions. Recall from Definition 1.1.4 that, given a set A, we say that the sets $S_1, S_2, \ldots, S_n$ form a partition of A, if the following conditions are satisfied:

1. $A = S_1 \cup S_2 \cup \cdots S_n$;
2. For any two different sets S_i and S_j, we have $S_i \cap S_j = \varnothing$.

Our next example is a more formal version of Example 5.18. We present it here to underscore that equivalence relations arise naturally from partitions.

■ **Example 5.21** Let $S_1, S_2, \ldots, S_n$ be a partition for a set A. Define R to be the relation "in the same set of the partition". That is, if $a, b \in A$, we have aRb when $a, b \in S_i$ for some of the elements in the partition (see Figure 5.6). The verification that this is an equivalence relation is straightforward.

■

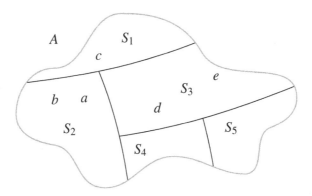

Figure 5.6: The sets S_1, S_2, S_3, S_4, and S_5 form a partition of the set A. The relation R is defined to mean "in the same set of the partition." In this case aRb, since a and b are in the same set S_2, and a is not related to d, as $a \in S_2$ and $d \in S_3$.

Definition 5.5.3 Let R be an equivalence relation on A. For any element $a \in A$, we can define the set

$$[a] = \{b \in A \mid aRb\}.$$

In other words, the set $[a]$ contains all elements from A that are related to a. The set $[a]$ is called the *equivalence class of a*.

The equivalence classes of a set A with an equivalence relation R defined on it have some important properties.

> **Theorem 5.5.1** Let R be an equivalence relation defined on a set A, and let $a, b \in A$ be such that aRb. Then $[a] = [b]$.

Proof. Let $a, b \in A$ and aRb. We will show that $[a] = [b]$ by showing that $[a] \subseteq [b]$ and $[b] \subseteq [a]$ (see Section 1.4.2).

Let $x \in [a]$, which means xRa. Since aRb, and the relation R is transitive, this means that xRb. That is $x \in [b]$. Therefore $[a] \subseteq [b]$

Now let $x \in [b]$, which means xRb. Since aRb, and the relation R is symmetric, we have that bRa. Now, since R is transitive, xRb and bRa show that xRa. Therefore $x \in [a]$, and we have proved that $[b] \subseteq [a]$. ∎

Our next result shows that many of the equivalence classes that correspond to the elements of A are identical.

> **Theorem 5.5.2** Let R be an equivalence relation defined on the set A and $a, c \in A$. If $[a] \cap [c] \neq \varnothing$, then $[a] = [c]$.

Proof. Since $[a] \cap [c] \neq \varnothing$, let $x \in [a] \cap [c]$. Since $x \in [a]$, we have xRa. Since $x \in [c]$, we have xRc, and (since R is symmetric) cRx. Since cRx and xRa, the transitivity property of R gives that cRa. From Theorem 5.5.1, we can now conclude that $[a] = [c]$. ∎

Theorem 5.5.2 shows that if we take the equivalence classes of any two elements of A, they are either the same or disjoint. We have thus proved a fundamental result about equivalence relations.

> **Theorem 5.5.3** The different equivalence classes for an equivalence relation R on a set A form a partition of A. That is, different equivalence classes do not overlap, and every element of the set A is in exactly one of the equivalence classes.

> **Exercise 5.24** Prove that each of the following is an equivalence relation. Then describe the partition associated with each relation.
> 1. The set $\mathbb{Z}$ with the relation aRb defined as "a and b have the same parity."

2. The set $\mathbb{Z}$ with the relation aRb defined as $|a| = |b|$.

3. The set $\mathbb{Q}$ with the relation aRb defined as $a - b \in \mathbb{Z}$.

4. Consider a set S and let $\mathscr{P}(S)$ be the set of all subsets of S. Let $D \subseteq S$. On $\mathscr{P}(S)$ the relation R is defined in the following way: two subsets A and B are related, when $A \cap D = B \cap D$.

5.5.2 Cosets

We now consider at a special type of partitions arising from subgroups.

> **Theorem 5.5.4** let G be a group and H be a subgroup of G. For each $x, y \in G$, consider the relation R on G defined in the following way: xRy means $xy^{-1} \in H$. Then R is an equivalence relation. The same is true for the relation R defined by xRy means $x^{-1}y \in H$.

Proof. We give the proof for xRy given by $xy^{-1} \in H$. The proof for xRy when $x^{-1}y \in H$ is left as an exercise (see Exercise 5.25). We have to check that R is reflexive, symmetric, and transitive.

(1) For any $x \in G$, we have $xx^{-1} = e \in H$. Thus R is reflexive.

(2) To show symmetry, we need to show that if xRy, then yRx. Let now xRy, which means $xy^{-1} \in H$. Since H is a subgroup, $yx^{-1} = (xy^{-1})^{-1} \in H$, which proves that yRx.

(3) To show that R is transitive, notice that if xRy and yRz, then $xy^{-1} \in H$ and $yz^{-1} \in H$. Since H is a subgroup, it is closed under multiplication, and $(xy^{-1})(yz^{-1}) = x(yy^{-1})z^{-1} = xz^{-1} \in H$, using the associativity property of groups. Thus xRz, proving that R is transitive. ∎

The equivalence relations from Theorem 5.25 give a partition of the group G with equivalence classes of special interest called cosets of G.

> **Definition 5.5.4** If H is a subgroup of G and $g \in G$ is fixed, then the set $Hg = \{hg \mid h \in H\}$ is called a *right coset* of H in G. The set $gH = \{gh \mid h \in H\}$ is called a *left coset* of H in G. If the group operation in G is addition, the right coset is denoted by $H + g = \{h + g \mid h \in H\}$ and the left coset is denoted by $g + H = \{g + h \mid h \in H\}$.

Note that in right cosets, the fixed element g is the right factor in the operation hg, hence the name. For left cosets, g is the left factor.

Theorem 5.5.5 If G is a group with a subgroup H, the right cosets Hg are the equivalence classes for the relation R on G with xRg defined as $xg^{-1} \in H$. Similarly, the left cosets gH are the equivalence classes for the relation gRy defined by $g^{-1}y \in H$.

Proof. We will prove the result for the right cosets of G. The proof regarding the left cosets is left as an exercise.

Let $g \in G$. Consider the set $S = \{x \in G \mid xg^{-1} \in H\}$. We will show that this set is the right coset Hg. We will establish this by showing that $S \subseteq H$ and that $H \subseteq S$.

First, let $x \in S$. Then, since $xg^{-1} \in H$, we will have $xg^{-1} = h$, for some $h \in H$, which we can rewrite as $x = hg \in Hg$. This shows $S \subseteq Hg$.

Next, let $x \in Hg$. This, by the definition of the right coset H, means that there is an $h \in H$ such that $x = hg$. But this is the same as $xg^{-1} \in H$, which shows that $Hg \subseteq S$. This proves that the equivalence classes of R are the right cosets Hg. ∎

With the information that the cosets form a partition for G, we now have the following important result.

Theorem 5.5.6 Let H be a subgroup of G. Then the left (right) cosets form a partition of G. Further, for finite groups, all cosets have the same number of elements.

Proof. Combining Theorems 5.5.3 and 5.5.5, we have a proof that the cosets of H in G form a partition of G. Thus, any two cosets are identical or disjoint. To see that all cosets have the same size, we will show that for any $a, b \in G$,

$$|Ha| = |H| = |bH|.$$

We will prove the first equality. The second one is is proved in a similar way and is left as an exercise (see Exercise 5.28).

Recall from Section 6.9.6 in the Appendix that two sets A and B have the same size, if there is a bijection $\phi : A \to B$.

Now let $a, b \in G$. Define a function $\phi : H \to Ha$ by $\phi(h) = ha$. We will show that ϕ is one-to-one and onto.

To check the one-to-one property, let $\phi(h_1) = \phi(h_2)$; that is, $h_1a = h_2a$. Multiplying by a^{-1} from the right gives $h_1 = h_2$.

To check that ϕ is onto, let $g \in Ha$. This means, there is an $h \in H$, such that $g = ha$. Thus, $g = \phi(h)$. ∎

The next fact plays a very important role in the theory of finite groups.

> **Theorem 5.5.7 Cayley's Theorem.** Let G be a finite group. If H is a subgroup of G, then the number of elements of H divides the number of elements of G. That is, $|H|$ divides $|G|$.

Proof. Let $H_1, H_2, \ldots, H_n$ be the distinct cosets of H. From Theorem 5.5.6, we know that the cosets form a partition and that they all have the same number of elements $|H|$. Thus

$$|G| = |\cup_{i=1}^n H_i| = |H_1| + |H_2| + \cdots + |H_n| = n|H|,$$

which means $|H|$ divides $|G|$, and $|H| = \frac{|G|}{n}$.

∎

> **Exercise 5.25** Prove that if G is a group with a subgroup H, the relation on G defined by xRy means $x^{-1}y \in H$ is an equivalence relation.

> **Exercise 5.26** Let G be a group with a subgroup H. Prove that if gRy is defined by $g^{-1}x \in H$, the equivalence classes for the relation R on G are the left cosets gH of H in G.

> **Exercise 5.27** Let G be a group with a subgroup H. Show that if $g \in H$, then $Hg = H = gH$.

> **Exercise 5.28** In the context of Theorem 5.5.6, prove that for any $b \in G$, $|H| = |bH|$.

> **Exercise 5.29** If G is a group of even order, prove that it has an element $a \neq e$ that satisfies $a^2 = e$.

5.6   Normal Subgroups and Quotient Groups

We saw in the examples that groups may have finite or infinite number of elements. Sometimes, even when the group is finite, its number of elements may be quite large. In this section, we discuss how the partition of a group given by its cosets can itself be turned into a group with an appropriate operation. That new group is smaller and preserves some of the structure of the original group. We begin with an example.

Recall from Example 5.14 that $n\mathbb{Z}$ is a subgroup of $\langle \mathbb{Z}, + \rangle$. In our example, we will use $n = 4$ to illustrate the main idea.

We consider a decomposition of $\mathbb{Z}$ into disjoint sets of the form $4\mathbb{Z} + r$, where $r = 0, 1, 2, \ldots, m - 1$. When $r = 0$ we get

$$4\mathbb{Z} = \{a \in \mathbb{Z} \mid a = 4m\} = \{\ldots, -8, -4, 0, 4, 8, 12, \ldots\}.$$

When $r = 1$,

$$4\mathbb{Z} + 1 = \{a \in \mathbb{Z} \mid a = 4m + 1\} = \{\ldots, -7, -3, 1, 5, 9, 13, \ldots\}.$$

Similarly, for $r = 2$ and $r = 3$, we get

$$4\mathbb{Z} + 2 = \{\ldots, -6, -2, 2, 7, 11, 14, \ldots\} \quad \text{and}$$
$$4\mathbb{Z} + 3 = \{\ldots, -5, -1, 3, 8, 12, 15, \ldots\}.$$

Now, when we take $r = 4$, we will get the same set as for $r = 0$:

$$4\mathbb{Z} + 4 = \{\ldots, -4, 0, 4, 8, 12, 16, \ldots\} = 4\mathbb{Z}.$$

In the same way, for $r = 5$, we get the set $4\mathbb{Z} + 1$; for $r = 6$, we get the set as $4\mathbb{Z} + 2$, and so on. Thus, we made an important observation: there is no point going beyond $r = 3$.

Another important observation is that the sets $4\mathbb{Z}, 4\mathbb{Z} + 1, 4\mathbb{Z} + 2$, and $4\mathbb{Z} + 3$ give a partition of $\mathbb{Z}$ (see Definition 1.1.4).

You have likely already noticed that:

1. Each of the sets $4\mathbb{Z}, 4\mathbb{Z} + 1, 4\mathbb{Z} + 2$, and $4\mathbb{Z} + 3$ is a coset of of the subgroup $4\mathbb{Z}$ in $\mathbb{Z}$, and they form a partition of $\mathbb{Z}$.
2. Each set $4\mathbb{Z} + r$, where $r = 0, 1, 2, 3$ gives us the set of all integers with a remainder r from division by 4.
3. Each set $4\mathbb{Z} + r$ for $r \geq 4$ is the set $4\mathbb{Z} + r_1$, where $r_1 = r \pmod 4$, so $0 \leq r_1 < 4$. Thus, there are only four different cosets.

An important thing about these cosets is that if we take any numbers from two of the cosets and add them together we always get a number that is in the same third coset. For example, if we take any number from the coset $4\mathbb{Z} + 1$ and add it to any number from the coset $4\mathbb{Z} + 2$, we always get a number in the coset $4\mathbb{Z} + 3$. If we take any number from the coset $4\mathbb{Z} + 2$ and add it to any number from the coset $4\mathbb{Z} + 3$, we always get a number in the coset $4\mathbb{Z} + 1$ due to the rules of modular arithmetic. Let's see why this is the case.

An element from $4\mathbb{Z}+2$ has the form $4n+2$ for some $n \in Z$. An element from $4\mathbb{Z}+3$ has the form $4m+3$ for some $m \in \mathbb{Z}$. Then

$$(4m+2)+(4n+3) = 4(m+n)+5 = 4(m+n+1)+1 = 1 \quad (\text{mod } 4).$$

This tell us that we can consider the set of four cosets $G = \{4\mathbb{Z}, 4\mathbb{Z}+1, 4\mathbb{Z}+2, 4\mathbb{Z}+3\}$ and turn it into a group with an operation +, defined by using the rules of modular arithmetic, namely:

$$(4\mathbb{Z}+2)+(4\mathbb{Z}+3) = 4\mathbb{Z}+5 = 4\mathbb{Z}+1, \quad \text{since } 5 \,(\text{mod } 4) = 1.$$

In general, if $a \neq b$ are two numbers in the set $\{0,1,2,3\}$, then we define the operation on G to be

$$(4\mathbb{Z}+a)+(4\mathbb{Z}+b) = 4\mathbb{Z}+r, \quad \text{where } r = a+b \,(\text{mod } 4).$$

If we remember that each element of G contains the $4\mathbb{Z}$ part, that is, if we *factor that part out*, we can think of G as $G = \{0,1,2,3\}$. And we already know that this is a group with the operation addition $(\text{mod } 4)$ (see Example 5.3)! Thus, we have found a way to break the group $\langle \mathbb{Z}, + \rangle$ into four disjoint cosets and form a new group, the elements of which are the cosets. The operation is addition $(\text{mod } 4)$.

This new group is called a *quotient group* and is denoted by $\mathbb{Z}/4\mathbb{Z}$. In your regular course in abstract algebra, you will prove that this quotient group behaves mathematically in the same way as the group $\mathbb{Z}_4$ [1].

In this example, everything worked out nicely because the set of cosets formed a group. It happened because we showed that if we take any numbers from two of the cosets and add them together we always get a number that is in the same third coset. This is not true for the set of cosets in general. Subgroups for which the cosets satisfy this property are called *normal subgroups*.

Definition 5.6.1 Let N be a subgroup of G. We say that N is a *normal subgroup* when

$$(Nx)(Ny) = N(xy), \quad \text{for all } x, y \in G. \tag{5.2}$$

If N is a normal subgroup of G, we write $N \lhd G$

In view of the example we considered above, this definition makes intuitive sense. However, we need to check that the choice of representative elements

---

[1] The term that is used to make this rigorous is that the quotient group $\mathbb{Z}/4\mathbb{Z}$ is *isomorphic* to $\mathbb{Z}_4$, but we will not be discussing it here.

$x, y \in G$ for the cosets Nx and Ny will not change the result of the operation. That is, we need to check that if x, y, x_1, and y_1 are any elements of G such that $Nx = Nx_1$ and $Ny = Ny_1$, then $N(xy) = N(x_1y_1)$. In mathematical language, this will mean that the operation is well defined.

So let $x_1, y_1 \in G$ be such that $Nx_1 = Nx$ and $Ny_1 = Ny$. Since $x_1 \in Nx$, this means, there is an element $n_1 \in N$ such that $x_1 = n_1x$. Similarly, there is an element $n_2 \in N$ such that $x_2 = n_2y$. Then

$$(Nx_1)(Ny_1) = (Nn_1x)(Nn_2y) = (Nx)(Ny) = N(xy),$$

because when n_1, $n_2 \in N$, $Nn_1 = N$ and $Nn_2 = N$ (see Exercise 5.27).

Another important property of normal subgroups is that for any $g \in G$, the left and the right coset of g are the same.

> **Theorem 5.6.1** If N is a normal subgroup of G, $gN = Ng$ for all $g \in G$.

Proof. Assume that N is a normal subgroup of G. Let $x, y \in G$ and $n_1, n_2 \in N$. Since N is a normal subgroup, we have $(Nx)(Ny) = N(xy)$. This means, that we can find $n_3 \in N$ such that

$$(n_1x)(n_2y) = n_3(xy).$$

Multiplying by y^{-1} from the right gives

$$n_1xn_2 = n_3x. \tag{5.3}$$

Multiplying both sides of Equation (5.3) by n_1^{-1} from the left gives

$$xn_2 = n_1^{-1}n_3x.$$

Now, since $n_1^{-1}n_3 \in N$, we have $xn_2 \in Nx$. Since $n_2 \in N$ was arbitrary, we have proved that $xN \subseteq Nx$.

Going back to Equation (5.3), we now multiply both sides by n_2^{-1} from the right, which leads to

$$n_1x = n_3xn_2^{-1}.$$

Since $xN \subseteq Nx$, there is a $n_4 \in N$ for which $xn_2^{-1} = n_4x$. So $n_1x = n_3n_4x \in Nx$. Because $n_1 \in N$ is arbitrary, this shows $Nx \subseteq xN$. Thus, we have proved that if N is a normal subgroup of G, $gN = Ng$ for all $g \in G$.

■

We just proved that for a normal subgroup the left and the right cosets are the same. This now allows us to rephrase Definition 5.6.1 by using left cosets the following way: A subgroup N of G is a normal subgroup, if the product of two left cosets xN and yN is the left coset $(xy)N$ for all $x, y \in G$.

An important comment is in order. If the group G is abelian, then $gH = Hg$ for any subgroup H. Thus all subgroups of an abelian group are normal subgroups. As our next example shows, for non-abelian groups, there may be subgroups H, for which $gH \neq Hg$ for some $g \in G$.

■ **Example 5.22** Let $G = S_3$, the symmetric group of 3 elements from Example 5.12. Consider the subgroup $H = \{\mu_1, \rho_0\}$. Then

$$\rho_1 H = \{\rho_1 \mu_1, \rho_1 \rho_0\} = \{\mu_3, \rho_1\},$$

while

$$H\rho_1 = \{\mu_1 \rho_1, \rho_0 \rho_1\} = \{\mu_2, \rho_1\} \neq \rho_1 H.$$

Thus, H is *not* a normal subgroup of S_3.   ■

We next show that the condition in Theorem 5.6.1 that, for a normal subgroup N, the left and right cosets are the same for all $g \in G$ is in fact an "if and only if" condition. That is, we will show next that N is a normal subgroup of G if an only if $Ng = gN$ for all $g \in G$. To streamline the proof, we will introduce some more terminology.

Definition 5.6.2 Let A, B be any two subsets of a group G. we define the product of the sets AB as

$$AB = \{ab \mid a \in A, \ b \in B\}.$$

Notice that, with this definition, the cosets of a subgroup H are products of sets where one of the sets has only a single element. We will prove a few properties the product of sets has.

Proposition 5.6.2 The product of sets is associative. That is, for any subsets A, B, C of G,

$$(AB)C = A(BC)$$

Proof. This follows from the associativity property of the group G.   ■

Proposition 5.6.3 If H is a subgroup of G, $HH = H$.

Proof. The proof is elementary, and we leave it as an exercise (see Exercise 5.31). ∎

We are now ready to state the main result.

> **Theorem 5.6.4** A subgroup N of G is a normal subgroup if and only if $Ng = gN$ for all $g \in G$.

Proof. We already established the "if-then" part in the proof of Theorem 5.6.1.

Now assume $Ng = gN$ for all $g \in G$. We will show that this implies $(Nx)(Ny) = N(xy)$ for all $x, y \in G$; that is, N is a normal subgroup by Definition 5.6.1.

Let $x, y \in G$. Now, using Propositions 5.6.2 and 5.6.3, we have

$$(Nx)(Ny) = N(x(Ny)) = N((xN)y) = N((Nx)y) = N(N(xy))$$
$$= (NN)(xy) = N(xy),$$

which proves N is a normal subgroup of G. ∎

There are other equivalent ways to check if a subgroup is normal. It is very common in mathematics to give a definition, then derive conditions that are equivalent to the definition but could be easier to check sometimes.

We now give several such criteria, equivalent to Definition 5.6.1.

> **Theorem 5.6.5** For N a subgroup of the group G, the following conditions are equivalent to N being a normal subgroup of G.
> 1. For every $g \in G$ and $n \in N$, $gng^{-1} \in N$;
> 2. For every $g \in G$, $gNg^{-1} \subseteq N$;
> 3. For every $g \in G$, $gNg^{-1} = N$;
> 4. For every $g \in G$, $gN = Ng$;
> 5. For every $x, y \in G$, $(Nx)(Ny) = N(xy)$.

Proof. We already know that 4 and 5 are equivalent. Proving that they are equivalent to the rest of the conditions can be accomplished by proving the following implications:

$$
\begin{array}{ccc}
1 & \Longleftarrow & 4 \\
\Big\Downarrow & & \Big\Uparrow \\
2 & \Longrightarrow & 3
\end{array}
$$

Proof of 4 ⇒ 1. Let $g \in G$ and $n \in N$. Since $gN = Ng$, there is a $n_1 \in N$ such that $gn = n_1 g$. Multiplying both sides by g^{-1} from the right, gives $gng^{-1} = n_1 \in N$.

Proof of 1 ⇒ 2. This is immediate since $gNg^{-1} = \{gng^{-1} \mid n \in N\}$.

Proof of 2 ⇒ 3. We need to show that for all $g \in G$, $N \subseteq gNg^{-1}$. We have $N = eNe^{-1} = (g^{-1}g)N(g^{-1}g) = g^{-1}(gNg^{-1})g \subseteq g^{-1}Ng$.

Proof of 3 ⇒ 4. Since $N = gNg^{-1}$ for all $g \in G$, the right coset $Ng = (gNg^{-1})g = gN(gg^{-1}) = gN$. ∎

Notice that it is essential to have the conditions of Theorem 5.6.5 be satisfied for all $g \in G$. As our next example shows, if H is a subgroup, it is possible to have $gHg^{-1} \subset H$ for *some* $g \in G$ and $gHg^{-1} \neq H$, so property 3 of Theorem 5.6.5 will not be satisfied.

■ **Example 5.23** Let's revisit Example 5.15. The group G is the group of 2×2 matrices with nonzero determinants with the operation matrix multiplication. We know that

$$H = \left\{ \begin{pmatrix} 1 & n \\ 0 & 1 \end{pmatrix} \,\middle|\, n \in \mathbb{Z} \right\}. \tag{5.4}$$

is a subgroup. Take the matrix

$$g = \begin{pmatrix} 2 & 0 \\ 0 & 1 \end{pmatrix} \in G, \text{ which has an inverse } g^{-1} = \begin{pmatrix} \frac{1}{2} & 0 \\ 0 & 1 \end{pmatrix}.$$

Now we have

$$gHg^{-1} = \left\{ \begin{pmatrix} 1 & 2n \\ 0 & 1 \end{pmatrix} \,\middle|\, n \in \mathbb{Z} \right\} \neq H$$

since gHg^{-1} is comprised only of matrices with even entries in the upper right corner, while H contains element that may have an odd number in that position. ■

Exercise 5.30 Show that the trivial subgroup $H = \{e\}$ is always a normal subgroup of any group G.

Exercise 5.31 Prove that if H is a subgroup of G, $HH = H$.

> **Exercise 5.32** Let H be a subgroup of G with the property that the product of every two right (left) cosets is a right (left) coset. Prove that $H \lhd G$. ∎

> **Exercise 5.33** The center of a group G is defined as
>
> $$Z(G) = \{g \in G \mid gx = xg, \text{ for all } x \in G\}.$$
>
> Prove that $Z(G)$ is a normal subgroup of G. ∎

Summary and What to Expect Next

Abstract algebra studies certain mathematical structures as abstract objects, which can then be instantiated in various contexts. Considered in the past to be a field of purely theoretical interest, abstract algebra now finds many applications in physics, chemistry, and engineering, as well as in cryptography, coding theory, and geometry. In this chapter, we introduced you to group theory – a topic that a full course in abstract algebra is likely to cover in more detail. Other abstract structures like rings, fields, and vector spaces can be seen as groups with additional structure introduced by adding operations and axioms, and your Abstract Algebra course will likely cover rings and fields.

At a higher level, certain extensions of group theory lead to the study of topological spaces and algebraic varieties. The latter are geometric objects defined as the set of solutions of systems of polynomial equations (and, as such, extend what you have learned about systems of linear equations). More generally, the field of algebraic geometry studies the geometry and structure of algebraic varieties. They are closely related to another algebraic structure – that of an ideal of a ring, where an ideal may be used to construct a quotient ring in a way conceptually similar to how normal subgroups can be used to construct quotient groups.

With reading and understanding this chapter, you have now made a confident first step on your way to study abstract algebra. We hope it will help you succeed in your traditional Abstract Algebra course, and we encourage you to continue the journey on your own afterward.

Suggested Videos and Further Reading

We highly recommend the excellent lessons on Abstract Algebra from the Socratica YouTube channel https://www.youtube.com/@Socratica dedicated to "math, science, and educational programming." The Socratica team has created a collection of lessons that can be used to supplement the material we have presented in this chapter, including groups, subgroups, Cayley tables, cosets, normal groups and quotient groups. We recommend that you watch each video before reading the relevant topic in the chapter, then watch it again after you have completed the reading and worked through the examples and exercises in the section.

We also recommend the following textbooks for further reading. They are standard texts used for undergraduate courses in Abstract Algebra.

Bibliography

[1] Lee, Gregory T. *Abstract algebra: An introductory course.* Springer, 2018.

[2] Pinter, Charles C. *A Book of Abstract Algebra.* Courier Corporation, 2010.

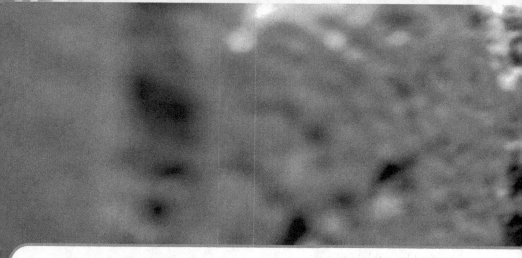

6. Appendix

Sets of Numbers

There are several number sets that are denoted by special symbols. We have collected them here.

Other common sets and the notation we use to denote them are:

$\mathbb{N} = \{1, 2, 3, ...\}$ – the set of natural numbers;

$\mathbb{Z} = \{..., -4, -3, -2, -1, 0, 1, 2, 3, ...\}$ – the set of integers;

$\mathbb{Q} = \{\frac{p}{q} \mid p \text{ and } q \text{ are integers}\}$ – the set of rational numbers;

$\mathbb{R} = \{x \mid x \in (-\infty, \infty)\}$ – the set of all real numbers.

Distance on the Real Line

Given a number $L \in \mathbb{R}$, the expression $|x - L|$ denotes the *distance* between the numbers x and L.

■ **Example 6.1** Let $L = 1$. Find all numbers x that satisfy $|x - L| = 2$. The expression $|x - L| = 2$ means that x is within distance 2 units from L. Moving two units to the left from $L = 1$ gives $x = -1$. Moving two units to the right of $L = 1$ gives $x = 3$ (see Figure 6.1). Thus, the values of x for which $|x - 1| = 2$, are $x = -1$ and $x = 3$. ■

DOI: 10.1201/9781032623849-6

Move 2 units to the left and right from $x = 1$

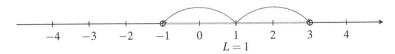

Figure 6.1: To find all x with $|x - 1| = 2$, use that $|x - 1|$ means x is at a distance two units from 1. The set of all x for which $|x - 1| < 2$ is the set of numbers within a distance from 1 that is less than 2; the interval $(-1, 3)$, marked by a thicker line.

■ **Example 6.2** For $L = 1$, find the set of all x that satisfy the inequality $|x - L| < 2$.

We are asking for the set of numbers x that are within distance from 1 that is smaller than 2. These are all numbers in the interval $(-1, 3)$ (see again Figure 6.1).   ■

■ **Example 6.3** Let $L \in \mathbb{R}$, $\varepsilon \in \mathbb{R}$, and $\varepsilon > 0$. Find the set of all numbers x that satisfy the inequality $|x - L| < \varepsilon$ and sketch it on the number line.

The inequality $|x - L| < \varepsilon$ means that the distance of x from L is smaller than ε. We draw a number line and mark the number L on it. We move a distance ε to the left of L, which takes us to the number $L - \varepsilon$. Moving the same ε distance to the write of L takes us to $L + \varepsilon$ (see Figure 6.2). The numbers $L - \varepsilon$ and $L + \varepsilon$ are at distance equal to ε from L. Thus, the numbers that are at a distance smaller than ε will be the numbers in the interval $(L - \varepsilon, L + \varepsilon)$, depicted by a thicker line.   ■

Move ε units to the left and to the right from the number L.

Figure 6.2: The numbers x for which $|x - L| < \varepsilon$ satisfy $(L - \varepsilon < x < L + \varepsilon)$ (see Exercise 6.3).

Example 6.3 allows us to make a very important observation that will be used repeatedly in Chapter 3:

The expression $|x - L| < \varepsilon$ is equivalent to $L - \varepsilon < x < L + \varepsilon$. (6.1)

Exercise 6.1 Solve the following equations for x:

1. $|x-4| = 5$;
2. $|x+1| = 7$ (*Hint.* First write $|x+1|$ as $|x-(-1)|$, which denotes the distance of x from the number -1.)
3. $|x-L| = 2$, where L is a fixed number.

Exercise 6.2 Solve the following inequalities for x:

1. $|x-4| > 5$;
2. $|x+1| \leq 7$ (*Hint.* First write $|x+1|$ as $|x-(-1)|$, which denotes the distance of x from the number -1.)
3. $|x-L| > 2$, where L is a fixed number.

The Triangle Inequality

The triangle inequality is a fundamental tool used in many proofs, including proofs in analysis and linear algebra.

Theorem 6.3.1 Let a and b be real numbers. Then

$$|a+b| \leq |a| + |b|.$$

Proof. We will use that for any number a, $|a| \geq a$ and $|a|^2 = a^2$. We have

$$
\begin{aligned}
(|a| + |b|)^2 &= |a|^2 + 2|a||b| + |b|^2 \\
&\geq a^2 + 2ab + b^2 \\
&= (a+b)^2.
\end{aligned}
$$

Taking square roots, gives

$$|a| + |b| \geq |a+b|.$$

∎

Factorial Notation

For any non-negative integer n, the notation $n!$ is used to denote the product of all positive integers that are less than or equal to n. More formally, we have

Definition 6.4.1 Let $n \geq 1$ be an integer. The notation $n!$ is used to denote the following product.

$$n! = 1 \cdot 2 \cdot 3 \cdot n \tag{6.2}$$

The notation $n!$ is pronounced "n-factorial."

■ **Example 6.4**

- $3! = 1 \cdot 2 \cdot 3 = 6$,
- $5! = 1 \cdot 2 \cdot 3 \cdot 4 \cdot 5 = 120$,
- $8! = 1 \cdot 2 \cdot 3 \cdot 4 \cdot 5 \cdot 6 \cdot 7 \cdot 8 = 40,320$.

                                                                                    ■

From the definition of factorial, we see that

$$n! = 1 \cdot 2 \cdot 3 \cdot \cdot (n-1) \cdot n = (n-1)! \cdot n, \text{ for } n \geq 1. \tag{6.3}$$

Thus, if we know $(n-1)!$, we can multiply by n to get $n!$. For instance, We know from the example above that $5! = 120$, so we can find

$$6! = 5! \cdot 6 = 120 \cdot 6 = 720.$$

Equation (6.3) is a convenient way for calculating factorials, and we want it to hold for all $n \geq 1$. But if we apply it to $n = 1$, we get that $1! = 0! \cdot 1$. Thus, Definition 6.4.1 is extended to $n = 0$, with

$$0! = 1. \tag{6.4}$$

6.5 Binomial Theorem

This theorem provides a formula for raising the sum of two real numbers a and b to an integer power $n \geq 1$.

Theorem 6.5.1 For any two $a, b \in \mathbb{R}$ and $n \in \mathbb{Z}$

$$(a+b)^n = \binom{n}{0}a^n + \binom{n}{1}a^{n-1}b + \binom{n}{2}a^{n-2}b^2 + \cdots + \binom{n}{n-1}ab^{n-1}$$
$$+ \binom{n}{n}b^n$$
$$= \sum_{k=0}^{n} \binom{n}{k}a^{n-k}b^k, \tag{6.5}$$

where

$$\binom{n}{k} = \frac{n!}{k!(n-k)!}, \text{ for } k = 0, 1, 2, \ldots, n.$$

Let's check what the theorem gives us in the familiar cases when $n = 2$ and $n = 3$.

We know that

$$(a+b)^2 = (a+b)(a+b) = a^2 + ab + ba + b^2 = a^2 + 2ab + b^2. \quad (6.6)$$

Now let's verify that Equation (6.5.1) in the Binomial theorem gives the same for $n = 2$. We have

$$\binom{2}{0} = \frac{2!}{0!2!} = 1, \quad \binom{2}{1} = \frac{2!}{1!1!} = 2, \quad \binom{2}{2} = \frac{2!}{2!0!} = 1.$$

Thus, from the Binomial theorem, we get

$$(a+b)^2 = \binom{2}{0}a^2b^0 + \binom{2}{1}ab + \binom{2}{2}a^0b^2 = a^2 + 2ab + b^2,$$

exactly as we calculated by foiling in Equation (6.6).

For $n = 3$,

$$\begin{aligned}
(a+b)^3 &= (a+b)^2(a+b) \\
&= (a^2 + 2ab + b^2)(a+b) \\
&= a^3 + a^2b + 2a^2b + 2ab^2 + b^2a + b^3 \\
&= a^3 + 3a^2 + 3ab^2 + b^3.
\end{aligned}$$

We leave it as an exercise for you to show that this result is in agreement with the statement of the Binomial theorems for $n = 3$.

Exercise 6.3 Apply the Binomial theorem to obtain that

$$(a+b)^3 = a^3 + 3a^2 + 3ab^2 + b^3.$$

Exercise 6.4 Apply the Binomial Theorem to show that if p and q are positive numbers with $p + q = 1$, then

$$\binom{n}{0}p^n + \binom{n}{1}p^{n-1}q + \binom{n}{2}p^{n-2}q^2 + \cdots + \binom{n}{n-1}pq^{n-1} + \binom{n}{n}q^n = 1.$$

6.6 The Fundamental Theorem of Arithmetic

Theorem 6.6.1 Every positive integer (except the number 1) can be represented in exactly one way, apart from rearrangement, as a product of one or more primes.

This important result is also sometimes called The Unique factorization theorem. Its proof is beyond the scope of these notes, but its statement should make sense from the examples we present.

■ **Example 6.5**

$$16 = 2^4$$
$$180 = 2^2 \cdot 3^2 \cdot 5$$
$$1823 = 1823 \qquad \text{The number 1823 is itself a prime.}$$
$$87450 = 2 \cdot 3 \cdot 5^2 \cdot 11 \cdot 53 \qquad \text{Double check!}$$

■

Once a factorization are arranged in the order of increasing prime factors, they are unique. This provides a convenient way of finding the prime number factorization of relatively small numbers "by hand."

■ **Example 6.6** Find the prime number factorization of 588.

This is an even number, so we begin by factoring out a 2, then continuing until we get an odd factor:

$$588 = 2 \cdot 294 = 2^2 \cdot 147.$$

We next try for factors of 3, since 3 is the next smallest prime. We get

$$588 = 2^2 \cdot 147 = 2^2 \cdot 3 \cdot 49.$$

Since 49 doesn't have factors of 3, we look at the next smallest prime, which is 5. As 49 does not have a factor of 5 either, we move to 7 to get the factorization

$$588 = 2^2 \cdot 3 \cdot 7^2.$$

If we want to have every prime number listed in the factorization in increasing order, we could write

$$588 = 2^2 \cdot 3 \cdot 5^0 \cdot 7^2.$$

∎

> **Exercise 6.5** Find the prime factorization for each integer. Arrange the prime factors in increasing order.
>   1. 54
>
>   2. 265
>
>   3. 1111
>
>   4. 7547

∎

Integer Division and Modular Arithmetic

We all know that if we take two real numbers $x, y \in \mathbb{R}$, their sum, difference, product, and quotient are also real numbers. We often describe this by saying that the set of real numbers $\mathbb{R}$ is *closed* with respect to these operations. However, if we consider the set of integers $\mathbb{Z}$, the sum, the difference, and the product of two integers will still produce an integer, while the quotient of two integers may not be an integer. Thus, the set $\mathbb{Z}$ is *not closed* with respect to the division operation.

This leads to a different way of looking at the division operation called *integer division* or *division with a remainder*. This operation is of special interest in the sub-field of mathematics called number theory. In this section, we will provide the minimal background needed to follow the material included in this book. For a more detailed introduction, you may consider a traditional textbook in discrete mathematics such as [2] or [1].

In Chapter 1 we talk about how important definitions are for constructing rigorous mathematical proofs. You should make an effort to understand what property each definition in this section formalizes, study the examples carefully, *know* the definition, and understand how to use it properly in a proof.

Definition 6.7.1 Given two integers a and b, we say that *a divides b*, if

$$b = ak, \text{ for some integer } k \in \mathbb{Z}.$$

We write $a \mid b$, which is read as "a divides b." When this is *not* the case, we write $a \nmid b$, which is read as "a does not divide b."

We also say that a *is a divisor for* b, that b *is divisible by* a, and that b *is a multiple of* a.

■ **Example 6.7** According to this definition, we have that:

1. $3 \mid 6$ (3 divides 6), because, for $k = 2$, we have $6 = 3 \cdot 2$;

2. $7 \mid 28$ (7 divides 28), because, for $k = 4$, we have $28 = 7 \cdot 4$;

3. $-100 \mid 1000$ (100 divides 1000), because, for $k = -10$, we have $1000 = 100 \cdot (-10)$;

4. $3 \nmid 13$ (3 does not divide 13), because there is no integer k for which $13 = 3k$;

5. $5 \nmid -72$ (5 does not divide -72), because there is no integer k for which $-72 = 5k$.

■

How do we decide if there is an integer k with the desired property? One can use regular division to decide. If

$$\frac{b}{a} = k \text{ is an integer, then clearly } b = ak, \text{ with } k \in \mathbb{Z}.$$

If

$$\frac{b}{a} \text{ is not an integer, then } a \nmid b.$$

■ **Example 6.8** Let's revisit Example 6.7, using regular division. We have:

1. $3 \mid 6$, because $\frac{6}{3} = 2$, and $k = 2$ is an integer;

2. $7 \mid 28$, because $\frac{28}{7} = 4$, and $k = 4$ is an integer;

3. $-100 \mid 1000$, because $\frac{1000}{-100} = -10$, and $k = -10$ is an integer;

4. $3 \nmid 13$, because $\frac{13}{3} = 4.\overline{3}$, and $k = 4.\overline{3}$ is not an integer;

5. $5 \nmid -72$, because $\frac{-72}{5} = -14.4$, and $k = -14.4$ is not an integer.

■

The next result is fundamental when using integer arithmetic. We will present it here without proof, but it can be found in any standard text in discrete mathematics (e.g., [2]). The examples that follow should be sufficient for you to understand the concept. We will refer to this result sometimes as *division with remainder*.

Theorem 6.7.1 (*The Division Algorithm*) For any two integers $a \in \mathbb{N}$ and $b \in \mathbb{Z}$ there are *unique* integers such that

$$b = aq + r, \text{ where } 0 \leq r < a. \tag{6.7}$$

The integer q is called the *quotient* for the division of b by a. The integer r is the *remainder*. Notice that we have to have r be *strictly smaller than a*.

■ **Example 6.9** Find the quotient and the remainder from the division algorithm for the integers below:

1. $b = 14, a = 3$.
   We have $14 = 3 \cdot 3 + 2$. Thus $q = 3$ and $r = 2$ (so the requirement $0 \leq r < a$ is satisfied).

2. $b = 89, a = 7$.
   We have $89 = 7 \cdot 12 + 5$. Thus $q = 12$ and $r = 5$ (so the requirement $0 \leq r < a$ is satisfied).

3. $b = -17, a = 3$.
   We have $-17 = 3 \cdot (-6) + 1$. Thus $q = -6$ and $r = 1$ (so the requirement $0 \leq r < a$ is satisfied).

4. $b = -177, a = 4$.
   We have $-177 = 4 \cdot (-45) + 3$. Thus $q = -45$ and $r = 3$ (so the requirement $0 \leq r < a$ is satisfied).

5. $b = -352, a = 4$.
   We have $352 = 44 \cdot 4$. Thus $q = -44$, and $r = 0$ (so the requirement $0 \leq r < a$ is satisfied).

■

You may wonder how we found q and r in the previous example. Once again, we can use regular division, considering the fraction $\frac{b}{a}$. Once we find the quotient q, the remainder r can be found as $r = b - aq$.

Finding the quotient q is done based on whether b is positive or negative.

- If $b \geq 0$, the quotient q is the whole part of the fraction $\frac{b}{a}$.
- If $b < 0$, the quotient q is obtained by rounding off $\frac{b}{a}$ *down* to the next negative integer.

■ **Example 6.10** We'll revisit Example 6.9, now using regular division to find q and r.

1. $b = 14, a = 3$.
   We have $\frac{b}{a} = \frac{14}{3} = 4.\bar{6}$. Thus $q = 4$ and $r = b - aq = 14 - 3 \cdot 4 = 2$.

2. $b = 89, a = 7$.
   We have $\frac{b}{a} = \frac{89}{7} = 12.7143$. Thus, $q = 12$ and $r = b - aq = 89 = 7 \cdot 14 = 5$.

3. $b = -17, a = 3$.
   We have $\frac{-17}{3} = -5.\bar{6}$. Thus $q = -6$, as -6 is the next integer *down* the number line from $-5.\bar{6}$. Then we find and $r = b - aq = -17 - 3 \cdot (-6) = 1$ (so the requirement $0 \le r < a$ is satisfied).

4. $b = -177, a = 4$.
   We have $\frac{-177}{4} = -44.25$. Thus $q = -45$, as -45 is the next integer *down* the number line from -44.25. Then we find and $r = b - aq = -177 - 4 \cdot (45) = 3$ (so the requirement $0 \le r < a$ is satisfied).

5. $b = -352, a = 4$.
   We have $\frac{b}{a} = \frac{-352}{4} = 44$. Since we got an integer, $q = -44$, and $r = 0$.

■

Notice that when $r = 0$, we have $b = aq$, with $q \in \mathbb{Z}$. Thus, when $r = 0$, we have that $a \mid b$. Notice also that when $r \ne 0$, the condition $0 < r < a$ implies that $a \nmid b$. Thus, we have that

$a \mid b$ if and only if, in the division algorithm, the remainder $r = 0$.

Since the integer remainder r is required to satisfy $0 \le r < a$, it can only take one of the values $0, 1, 2, \ldots, a - 1$. When $a = 2$, the possible values are $r = 0$ and $r = 1$. Thus, the set of all integers can be split into two subsets – the subset of integers, for which 2 is a divisor, and the subset of integers that give remainder 1 when divided by $a = 2$. These are the well-known subsets of even and odd numbers.

Definition 6.7.2 An integer b is *even* if it can be written as

$b = 2q$, for some $q \in \mathbb{Z}$.

Definition 6.7.3 An integer b is *odd* if it can be written as

$$b = 2q + 1, \text{ for some } q \in \mathbb{Z}.$$

Very often we are not interested in the exact value of a number but interested in its parity.

Definition 6.7.4 The classification of a number as even or odd is referred to as its *parity*.

■ **Example 6.11** Determine if the numbers have the same parity.

1. $a = -98, b = 13$;
   We have $a = -98 = 2 \cdot (-49)$, which shows that -98 is even. On the other hand, $b = 13 = 2 \cdot 6 + 1$, so 13 is odd. Thus, the numbers have different parity.

2. $a = 765, b = 17$;
   We have $a = 765 = 2 \cdot (384) + 1$, which shows that 765 is odd. Next, $b = 17 = 2 \cdot 8 + 1$, so 17 is also odd. Thus, the numbers have the same parity.

                                                                          ■

The set of integers that give the same remainder when divided by a fixed positive integer m often share properties of interest. For instance, that the sum of any two odd integers is an even integer (see Example 1.17 for the proof). Or that the sum of any number a that gives reminder $r = 1$ when divided by $m = 3$ and any number b that gives reminder $r = 2$ when divided by the same integer $m = 3$ is a multiple of 3 (see Example 1.19 for the proof). This justifies the use of special terminology.

Definition 6.7.5 Let a, b, and $m > 0$ be integers. If a and b give the same remainder when divided by m we write

$$a = b \pmod{m},$$

and we say that *a and b are congruent* mod m.

Notice that when $a = mq + r$, with $q, r \in \mathbb{Z}$ and $0 \le r < m$, this definition implies

$$a = r \pmod{m},$$

since $r = m \cdot 0 + r$.

■ **Example 6.12** Identify each of the statements as True or False and explain.

1. $3 = 5 \pmod 2$;
   The statement is true since $3 = 2 \cdot 1 + 1$ and $5 = 2 \cdot 2 + 1$. Thus, both numbers give remainder $r = 1$ when divided by 2.

2. $78 = 16 \pmod 4$;
   The statement is false since $78 = 4 \cdot 19 + 2$ and $16 = 4 \cdot 4$. Thus, 78 gives remainder 2 when divided by 4, while the remainder of 16 when divided by 4 is zero.

3. $-177 = 19 \pmod 4$;
   The statement is true since both -177 and 19 give the same remainder $(r = 3)$ when divided by 4 (see Example 6.10).

4. $689 = 4 \pmod 5$.
   The statement is true since $689 = 5 \cdot 137 + 4$ and $4 = 5 \cdot 0 + 4$, so both remainders from dividing by 5 are $r = 4$.

■

Definition 6.7.5 allows us to consider integers congruent $\pmod m$ as equivalent in the sense that we only pay attention to their remainder and ignore the quotient. In Chapter 5, we use the following notation.

Given a positive integer m, we define the sets

$$m\mathbb{Z} = \{a \in \mathbb{Z} \mid a = mq, \text{ for some } q \in \mathbb{Z}\};$$
$$m\mathbb{Z} + 1 = \{a \in \mathbb{Z} \mid a = mq + 1, \text{ for some } q \in \mathbb{Z}\};$$
$$m\mathbb{Z} + 2 = \{a \in \mathbb{Z} \mid a = mq + 2, \text{ for some } q \in \mathbb{Z}\};$$
$$\vdots$$
$$m\mathbb{Z} + (m - 1) = \{a \in \mathbb{Z} \mid a = mq + (m - 1), \text{ for some } q \in \mathbb{Z}\}.$$

Sometimes, we will also equate any two integers from the same set and say that they are equal *equal* $\pmod m$.

■ **Example 6.13** With this notation, we have that:
1. $2\mathbb{Z}$ is the set of all even integers. Thus, if $a \in 2\mathbb{Z}$, $a = 0 \pmod m$;
2. $2\mathbb{Z} + 1$ is the set of all odd integers. Thus, if $a \in 2\mathbb{Z} + 1$, $a = 1 \pmod m$;
3. $7\mathbb{Z} + 4$ is the set of all integers that give remainder 4 when divided by 7. Thus, when $a \in 7\mathbb{Z} + 4$, $a = 7q + 4$ for some $q \in \mathbb{Z}$ and $a = 4 \pmod 7$.

■

What happens if we consider the sets $m\mathbb{Z}+m, m\mathbb{Z}+(m+1), m\mathbb{Z}+(m+2)$, and so on?

Because we only have m different remainders from division by m ($r = 1, 2, \ldots, m-1$), if we consider the set $m\mathbb{Z}+m$, we will have the set of integers of the form $a = mq+m$ for some $q \in \mathbb{Z}$. Because $a = mq+m = m(q+1) = mk$ for some $k \in \mathbb{Z}$, this gives the same set as $m\mathbb{Z}$.

In the same way, we can see that $m\mathbb{Z}+(m+1) = m\mathbb{Z}+1, m\mathbb{Z}+(m+2) = m\mathbb{Z}+2$, and, in general,

$$m\mathbb{Z}+n = m\mathbb{Z}+r, \quad \text{where } r = n \pmod{m}.$$

Thus, there are only m different sets of the form $m\mathbb{Z}+n, n \in \mathbb{Z}$.

We have thus arrived at a very important result: if we convert to $\pmod{m}$ after adding two integers, we will get the same result as adding the $\pmod{m}$ congruents for each integer first and then taking the result $\pmod{m}$.

> **Theorem 6.7.2** Let $a, b \in \mathbb{Z}$. Then
>
> $$(a+b) \pmod{m} = (a \pmod{m}) + (b \pmod{m}) \pmod{m}.$$

Functions

The concept of a function is fundamental in mathematics. It defines a special relation between the elements of two sets X and Y. We usually think of one of the sets as the set of inputs and the other as the set of outputs. The notation

$$f : X \to Y$$

identifies X as the set of inputs and Y as the set of outputs. The rigorous mathematical definition is this.

> **Definition 6.8.1** We say that the mapping $f : X \to Y$ between two sets X and Y is a *function* if
> 1. Every element $x \in X$ is mapped to an element $y \in Y$, and
> 2. For each $x \in X$, the corresponding element $y \in Y$ is unique.
>
> The unique element y for each x is denoted by $y = f(x)$.
>
> The set X is called the *domain* of the function f and Y is called the *co-domain* of f.

Figure 6.3 presents a simple illustration of mappings $f : X \to Y$ for $X = \{a, b, c\}$ to the set $Y = \{u, v\}$. The correspondence in panel (a) depicts a function, as both conditions from Definition 6.8.1 are satisfied. The correspondence in panel (b) is not a function (the element $c \in X$ does not have an

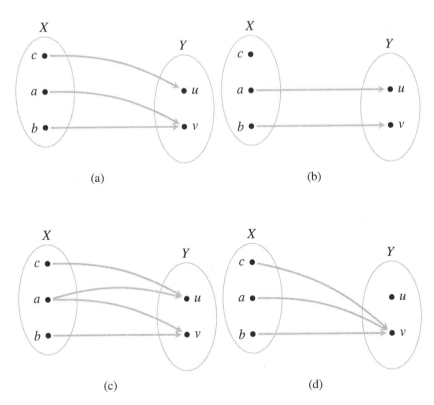

Figure 6.3: The mappings from $X = \{a, b, c\}$ to $Y = \{u, v\}$ depicted in panels (a) and (d) are functions while those in panels (b) and (c) are not.

output), neither is mapping in panel (c) (the element $a \in X$ has two outputs). Finally, the correspondence presented in panel (d) depicts a function, as all conditions in Definition 6.8.1 are satisfied.

6.9   Functions with Special Properties

Note that Definition 6.8.1 does *not* stipulate that the outputs for different $x \in X$ should be different. The functions in panels (a) and (d) of Figure 6.3 map a and b to the same element $v \in Y$, and this does not violate either condition in the definition. The subclass of functions we define next require that this not happen.

Definition 6.9.1 The function $f : X \to Y$ is called *one-to-one* when for any $x_1 \neq x_2$ in X, we have $f(x_1) \neq f(x_2)$.

In Chapter 1, we prove that the following definition is equivalent (see Theorem 1.2.2).

Definition 6.9.2 The function $f : X \to Y$ is called *one-to-one* when $f(x_1) = f(x_2)$ implies $x_1 = x_2$, for any $x_1, x_2 \in X$.

■ **Example 6.14** Figure 6.4 depicts a one-to-one function from $X = \{a, b, c\}$ to $Y = \{u, v, w, t\}$. ■

■ **Example 6.15** The function $f : \mathbb{R} \to \mathbb{R}$ with $f(x) = x^3$ is a one-to-one function defined over the entire real line: $x_1 \neq x_2$ implies $x_1^3 \neq x_2^3$ for all $x, x_2 \in \mathbb{R}$. ■

■ **Example 6.16** The function $f : \mathbb{R} \to \mathbb{R}$ with $f(x) = x^2$ is not one-to-one because, e.g., $f(2) = 4 = f(-2)$ while $2 \neq -2$. ■

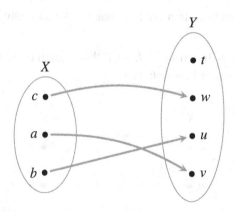

Figure 6.4: A one-to-one function from X to Y.

Note further that for f to be a function, not every element of Y has to be an image of some $x \in X$. Only a special class of functions have that property.

Definition 6.9.3 We say that a function $f : X \to Y$ is *onto* if for every $y \in Y$, there is an $x \in X$, such that $y = f(x)$.

The function in Figure 6.3, panel (a) is onto, while the function in panel (d) is not (there is no $x \in X$ with $u = f(x)$).

■ **Example 6.17** The function $f : \mathbb{R} \to \mathbb{R}_+$ with $f(x) = e^x$ is onto. For any $y \in \mathbb{R}_+$, the value $x = \ln y \in \mathbb{R}$ satisfies $e^x = e^{\ln y} = y$. If we consider the same function f with $f : \mathbb{R} \to \mathbb{R}$, now f is not onto: if $y \in \mathbb{R}$ is negative, there is no $x \in \mathbb{R}$ for which $y = e^x$. ■

■ **Example 6.18** The function $f : \mathbb{R} \to \mathbb{R}$ defined as $f(x) = \sin x$ is not onto. If $y \notin [-1, 1]$, no value of x will have $\sin(x) = y$. However, the same function with codomain $[-1, 1]$, is onto (and gives another example of a function that is not one-to-one). ■

When a function is not onto, it is of interest to identify the subset of elements of Y that are images of the elements in X.

Definition 6.9.4 Let $f : X \to Y$ be a function. The subset $Z \subset Y$ such that for any $z \in Z$, there is an $x \in X$ with $z = f(x)$ is called the *range* of the function f.

■ **Example 6.19** The range of the function in Figure 6.3, panel (a) is the entire set $Y = \{u, v\}$. The range of the function in panel (d) is $Z = \{v\}$. The range of the function in Figure 6.4 is $Z = \{u, v, w\}$. ■

The class of functions that are both one-to-one and onto is often of special importance.

Definition 6.9.5 A function $f : X \to Y$ that is both one-to-one and onto is called a *bijection* (see Figure 6.5).

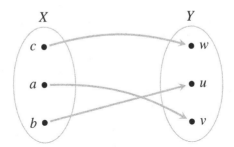

Figure 6.5: A bijection from X to Y.

■ **Example 6.20** The function $f : \mathbb{R} \to \mathbb{R}$ defined as $f(x) = x^3$ is a bijection. ■

■ **Example 6.21** The function $f : [-\frac{\pi}{2}, \frac{\pi}{2}] \to [-1, 1]$ defined as $f(x) = \sin x$ is a bijection. ■

Notice that when the sets X and Y are finite, having a function $f : X \to Y$ that is a bijection implies that X and Y have the same number of elements. For infinite sets, we use this property to check if two sets have the same size.

Definition 6.9.6 We say that two sets A and B *have the same size* if there exists a bijection $f : A \to B$.

This definition prompts us to think about the "size" of infinite sets. We already know that some infinite sets are larger than others in the sense that one is a strict subset of the other. Examples include $\mathbb{Z} \subset \mathbb{Q}$, $N \subset \mathbb{Z}$, $\mathbb{Q} \subset \mathbb{R}$. However, the definition above allows us to decide when various (and sometimes rather complex) infinite sets have the same infinite size.

Composition of Functions

If the output of one function is used as an input for another function, we talk about *composing* the two functions.

If $f : X \to Y$ and $g : Y \to Z$ be functions, and let the range of f be Y (that, is, f is onto). Then, we can construct a new function $h : X \to Z$, by defining $h(x) = (g \circ f)(x)$ as follows:

$$h(x) = (g \circ f)(x) = g(f(x)).$$

In other words, given $x \in X$, we apply f first and use its output $f(x)$ as an input for g. The function h is the *composition* of the functions f and g.

■ **Example 6.22** Consider $f : \mathbb{R} \to \mathbb{R}$ defined as $f(x) = x^3 + 5$ and $g : \mathbb{R} \to \mathbb{R}$ be $g(x) = x^2$. Now

$$h(x) = (g \circ f)(x) = g(f(x) = g(x^3 + 5) = (x^3 + 5)^2.$$

■

Notice that, in general, $(g \circ f)(x) \neq (f \circ g)(x)$.

■ **Example 6.23** For the functions from Example 6.22, let's calculate $(f \circ g)(x)$. We have

$$(f \circ g)(x) = f(g(x)) = f(x^2) = (x^2)^3 + 5 = x^6 + 5 \neq (x^3 + 5)^2 = (g \circ f)(x).$$

■

■ **Example 6.24** Let $f(x) = \sin x$ and $g = x^2 + 2x - 1$. We now have

$$(f \circ g)(x) = f(g(x)) = f(x^2 + 2x - 1) = \sin(x^2 + 2x - 1).$$

Composing the functions in reverse order gives

$$(g \circ f)(x) = g(f(x)) = g(\sin x) = (\sin x)^2 + 2\sin x - 1.$$

Notice again that $g(f(x)) \neq f(g(x))$.

■

6.11 Inverse Functions

When a function f, $f : X \to Y$ is a bijection, we can define a function $g : Y \to X$ that "undoes" the operation of f.

> **Definition 6.11.1** Let $f : X \to Y$ be a bijection. The function $g : Y \to X$ is called *the inverse function of* f, if
> 1. for every $x \in X$, $(g \circ f)(x) = g(f(x)) = x$, and
> 2. for every $y \in Y$, $(f \circ g)(y) = f(g(y)) = y$.
>
> The inverse g with these properties is denoted by f^{-1}, which is read as "f inverse."

Notice that since f^{-1} is defined on the set Y, we should write $f^{-1} = f^{-1}(y)$. However, since it is customary to denote the independent variable by x, we usually give inverse functions as functions of x, keeping in mind that x should be in the domain of f^{-1}.

We next explain by examples how to find f^{-1} for a bijection f. Each of the examples highlights the steps one follows to determine the inverse of a real-valued function.

■ **Example 6.25** Let $f(x) = 3x + 7$. Find $f^{-1}(x)$.

Step 1. Set $y = 3x + 7$;

Step 2. Solve for x. We obtain

$$x = \frac{y-7}{3}.$$

Step 3. Now, we have

$$f^{-1}(y) = \frac{y-7}{3}.$$

Because we usually denote the independent variable of a function by x, the function $f^{-1}(x)$ is thus

$$f^{-1}(x) = \frac{x-7}{3}.$$

Let's now check that f^{-1} satisfies the properties from Definition 6.11.1. We have

$$f^{-1}(f(x)) = f^{-1}(3x+7) = \frac{(3x+7)-7}{3} = \frac{3x}{3} = x,$$

$$f(f^{-1}(x)) = f\left(\frac{x-7}{3}\right) = 3\left(\frac{x-7}{3}\right) + 7 = (x-7)+7 = x.$$

We have now verified that $f^{-1}(x) = \frac{x-7}{3}$.   ■

■ **Example 6.26** Let $f(x) = -2x + 12$. Find $f^{-1}(x)$.
 Step 1. Set $y = -2x + 12$;
 Step 2. Solve for x. We obtain

$$x = -\frac{1}{2}y + 6.$$

Step 3. Now, we have

$$f^{-1}(y) = -\frac{1}{2}y + 6.$$

Renaming the independent variable as x, gives us

$$f^{-1}(x) = -\frac{1}{2}x + 6.$$

                                                                                ■

■ **Example 6.27** Let $f(x) = x^3 + 8$. Find $f^{-1}(x)$.
 Step 1. Set $y = x^3 + 8$;
 Step 2. Solve for x. We obtain

$$x^3 = y - 8, \text{ thus } x = \sqrt[3]{y - 8}.$$

Step 3. We now have

$$f^{-1}(y) = \sqrt[3]{y - 8}.$$

Renaming the independent variable as x, gives us

$$f^{-1}(x) = \sqrt[3]{x - 8}.$$

                                                                                ■

Exercise 6.6 Check that the function f^{-1} we found in Exercise 6.26 satisfies the properties from Definition 6.11.1.   ■

Exercise 6.7 Check that the function f^{-1} we found in Exercise 6.27 satisfies the properties from Definition 6.11.1.   ■

6.12 Images and Preimages of Sets

Let $f : X \to Y$ be a function. Suppose A is a subset of X and B is a subset of Y. It is sometimes of importance to determine the set comprised of the images $f(x)$ for all $x \in A$ and the set of all $x \in X$ such that $f(x) \in B$.

Definition 6.12.1 Let $f : X \to Y$ be a function and $A \subset X$. The set

$$f(A) = \{y \in Y \mid f(x) = y, \text{ for some element } x \in A\}$$

is called the *image of the set A* under the function f.

■ **Example 6.28** Let $f(x) = x^2$. Take $A = [-2, 2]$. Then $f(A) = [0, 4]$, since $[0, 4]$ contains all y-values, such that $x^2 = y$, for some $x \in [-2, 2]$. If we consider the set of two points $S = \{-2, 2\}$, then $f(S) = \{4\}$. ■

■ **Example 6.29** Let $f(x) = \cos x$ and $A = [0, \frac{\pi}{2}]$. Then $f(A) = [0, 1]$, since any $y \in [0, 1]$ has a value $x \in [0, \frac{\pi}{2}]$, for which $y = \cos x$. ■

Definition 6.12.2 Let $f : X \to Y$ be a function and $B \subset Y$. The set

$$f^{-1}(B) = \{x \in X \mid f(x) \in B\}$$

is called the *preimage of the set B* under the function f.

■ **Example 6.30** Let $f(x) = x^2$ and $B = [0, 1]$, then the preimage $f^{-1}(B) = [-1, 1]$, since the values of x in $[-1, 1]$ are those that produce y-values in $[0, 1]$. ■

■ **Example 6.31** Let $f(x) = \sin x$ and $B = 0$. Now $f^{-1}(B) = \{x = \pi k \mid k \in \mathbb{Z}\}$ because $\sin(\pi k) = 0$ for all integer k, and these are the only values $x \in \mathbb{R}$, for which $\sin x = 0$. ■

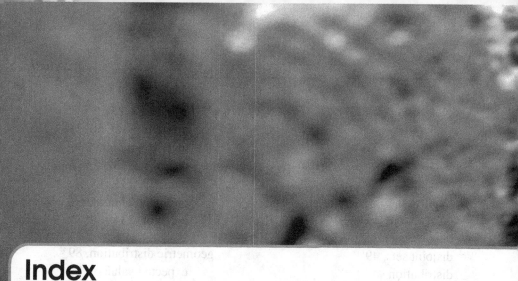

Index

Printed in the United States
by Baker & Taylor Publisher Services